Library of
Davidson College

WAGE CONTROL AND INFLATION
IN THE SOVIET BLOC COUNTRIES

By the same author

WAGE, PRICE AND TAXATION POLICY IN
CZECHOSLOVAKIA, 1948–70

WAGE CONTROL AND INFLATION IN THE SOVIET BLOC COUNTRIES

Jan Adam

© Jan Adam 1979

All rights reserved. No part of this publication may be reproduced or transmitted, in any form or by any means, without permission

First published 1979 by
THE MACMILLAN PRESS LTD
London and Basingstoke
Associated companies in Delhi
Dublin Hong Kong Johannesburg Lagos
Melbourne New York Singapore Tokyo

Reproduced from copy supplied
printed and bound in Great Britain
by Billing and Sons Limited
Guildford, London, Oxford, Worcester

British Library Cataloguing in Publication Data

Adam, Jan
    Wage control and inflation in the Soviet bloc countries
    1. Inflation (Finance)—Europe, Eastern
    2. Inflation (Finance)—Russia 3. Europe, Eastern—Economic policy 4. Russia—Economic policy
    I. Title
    332.4'1   HG925

    ISBN 0–333–27236–6

This book is sold subject to the standard conditions of the Net Book Agreement

To Zuzana and Julie

Contents

List of Tables	x
Acknowledgements	xi
Abbreviations	xiii
Introduction	xiv

PART ONE

1 PRICE STABILITY POLICIES IN THE SOVIET BLOC COUNTRIES — 3
    Soviet-type price system — 4
    Inflation in the initial period of economic planning — 6
    Price stabilisation policies — 8
    Diverse price stability policies — 17

2 WAGES AND INFLATION—A MACROECONOMIC ANALYSIS — 33
    Wages and productivity — 34
    Wages and wage plan targets — 45

PART TWO

3 REGULATION OF BASIC WAGES — 57
    Regulation of wage rates — 60
    Regulation of basic wages — 63
    Bank's control function — 72
    Collective agreements — 74

Contents

4	**REGULATION OF BONUSES**	77
	Formation and regulation of the bonus fund	79
	Incentive system for top managers	85
	The share of bonuses in total average wages	88
5	**WAGE CONTROL AND SUCCESS INDICATORS**	92
	Profit versus gross income—survey of opinions	93
	Indicators for wage control	102
6	**UNIFORMITY VERSUS DIFFERENTIATION**	108
	Uniform or differentiated taxation?	111
	Uniform or differentiated long-term normatives?	115

PART THREE

7	**WAGE REGULATION IN THE USSR (INCLUDING THE GDR)**	123
	The present system	123
	The incentive system of 1965	127
	Experiments in wage regulation	130
	Bank control	132
	Wage regulation in the GDR	134
8	**REGULATION OF WAGES IN POLAND**	137
	Reform in the 1960s	137
	The abortive reform of 1970	139
	The reform of 1973	141
	Modifications of 1976–7	146
9	**WAGE REGULATION IN HUNGARY**	151
	The reform of 1968	151
	Modifications in 1971	154
	Modifications of 1976	157
10	**WAGE REGULATION IN CZECHOSLOVAKIA**	164
	The reform of 1958	164
	The reform of 1966	165
	The present system	170

PART FOUR

11 **THE EFFECTIVENESS OF THE WAGE REGULATION SYSTEMS AS ANTI-INFLATIONARY TOOLS**   177
    Systems of management and inflation   179
    The effectiveness of wage regulation systems   181

Conclusion   193

Notes   196

Appendixes   229

Selected Bibliography   232

Index   238

List of Tables

1.1	Evolution of retail prices	6
1.2	Currency reforms	13
1.3	Price reductions	16
1.4	Cost of living	18
1.5	Retail price index in the GDR	20
1.6	Turnover tax and subsidies on consumer goods in 1969 in Hungary	21
1.7	Taxation of consumer goods in Czechoslovakia	22
1.8	Relationship of accumulation to costs of production of some consumer goods in Poland	23
2.1	Social labour productivity and average money wages	37
2.2	National income per employee and average money wages	38
2.3	Real wages	44
2.4	National income per capita and disposable money income per capita	45
2.5	Planned and actual average wage increases per wage earner	47
2.6	Employment	50
4.1	Bonuses in industry as percentages of total average wages in 1960–73	89

Acknowledgments

I would like to express my gratitude to the Social Sciences and Humanities Research Council of Canada for the extended research grants which enabled me to work on this study. My gratitude is also due to the University of Calgary for granting me the Killam Resident Fellowship which made it possible for me to expedite the completion of this book.

I am obliged to those who read the original draft, in part or whole, and whose comments enabled me to improve the final version of the book at many points. I would like to pay tribute especially to Professors W. Brus, D. Granick, M. Kaser and F. Levčik. I am also indebted to my colleagues at the University of Calgary, Professors S. Peitchinis and C. van de Panne. Of course, the sole responsibility for any remaining errors is mine.

Most of the materials for this book were collected during my sabbatical leave 1974–5 and my study stay in 1977 in Europe. My thanks are due to the Institute für Internationale Wirtschaftsvergleiche in Vienna, Osteuropa Institut in Munich and St Antony's College in Oxford for providing me with office space and other facilities during my stay there. In Europe I consulted many scholars and drew materials from many libraries and Institutes; I would like to express my gratitude especially to Radio Free Europe, Radio Liberty and Osteuropa Institut, all in Munich, Weltwirtschaft Institut in Kiel, Bundesinstitut für Ostwissenschaftliche und Internationale Studien in Cologne, and Deutsches Institut für Wirtschaftsforschung in Berlin, as well as to some of the researchers and employees of those institutes.

I also wish to record my appreciation of the help contributed by my research-assistant Dr L. Kivisild in collecting, processing, and evaluating materials as well as of the care and patience of Mrs B. Blackman who improved the English of my manuscript. To Mrs P. Dalgetty and S. Langan I am obliged for their care in typing the several drafts of the study.

A small part of the material in the study was published earlier in four articles: 'The System of Wage Regulation in Hungary', *Canadian Journal of Economics*, Vol. VII, no. 4, University of Toronto Press, 1974, pp. 578–93; 'The Recent Reform of the Incentive System in Poland', *Osteuropa Wirtschaft*, Vol. XX, no. 3, 1975, pp. 169–83 (this article also appeared in the research reports of the Institut für Internationale Wirtschaftsvergleiche in Vienna, no. 28, August 1975); 'Systems of the Wage Regulation in the Soviet Bloc', *Soviet Studies*, Vol. XXVIII, no. 1, January 1976, pp. 91–109 and 'Formation and Regulation of the Incentive Funds in the Soviet Bloc', in *Pioneering Economics*, International Essays in Honour of Professor Demaria, edited by T. Bagiotti and G. Franco, Padova, Italy, 1978, pp. 3–21. I wish to thank the editors of the above mentioned publications for permission to use the material contained in my articles in this book.

Finally, to my wife, Zuzana who encouraged me in my work and helped me to collect and process materials, I am indebted more than I can say.

Abbreviations

Full references to books are listed in the Bibliography. References in Notes indicate only the author of the book, year of its publication and page number.

References to periodical and newspaper articles are listed in the notes without the title of the article. For space reasons only titles of articles published in English are listed in the Bibliography.

The following abbreviations are used:

- **GDR** German Democratic Republic
- **NIE** national income per employee
- **SEIR** system of employment income regulation
- **SLP** social labour productivity
- **SWR** system of wage regulation
- **TUs** Trade unions

Introduction

This book deals with wage control[1] and inflation in the following five most important countries of the Soviet bloc—the USSR, Poland, the GDR, Czechoslovakia and Hungary. It has a twofold purpose: to analyse the role wages play in generating inflation (open and repressed), and, above all, to examine the methods the Soviet bloc countries use for and their effectiveness in controlling wage growth at the level of profit making (*khozraschet*) enterprises, primarily in industry. Since wage control aims not only at protecting the economy against inflationary pressures, but at least as much at improving the efficiency of the economy, the study will devote appropriate attention to that aspect too.

The book consists of four parts and a conclusion. Part One is made up of two chapters, the first of which serves in a sense as an expanded introduction. It briefly surveys the development of consumer prices and the methods used to achieve price stability since the start of economic planning. Also, it shows how market equilibrium was influenced by retail price distortions—a primary result of the adopted concept of prices—and by the desire to bring incomes in agriculture closer to the level of the non-agricultural sectors.

In the second chapter, an attempt is made to examine the role of wages in generating inflation (open and repressed) on the macroeconomic level. For this purpose, wage growth is compared with productivity growth as well as wage increases with wage plan targets. In the latter comparison, besides the role of average wages, the role of employment in the overexpenditure of wage increase targets is examined.

In the four chapters of Part Two the methods of wage control are discussed on a general level and in a comparative way. Chapter 3 classifies the methods used for wage regulation into three systems and examines their similarities and differences as subsystems of the systems of management of the economy. Special attention is devoted to the differences in the regulation of—what is the most important component of wage regulation—the growth of the basic wage-bill or basic average wages. Chapter 4 discusses the methods of regulating bonus funds in light of systemic differences. Since bonuses are earmarked largely for managers, the incentive system for top managers is also discussed. Chapter 5 deals with the choice of indicators. Since, during the reforms of the 1960s, all the countries except the USSR considered profit versus gross income or vice versa, a survey of views on the advantages and disadvantages of the two indicators is given. Chapter 6 analyses the problem of uniformity versus differentiation in the application of the tools of wage regulation, primarily with regard to taxation and long-term normatives.

Part Three consists of four chapters in which the system of wage regulation in individual countries is discussed. Since, in the GDR, the system has changed little in the last two decades, it is discussed in the framework of the chapter on the USSR. In these four chapters, the focus is directed to the above-mentioned most important component of wage control. Though the main stress of the analysis is on current methods, nevertheless we also focus our attention on the past, as far as methods different from the classical (Soviet) were used. Part Three, taken as a whole, indirectly gives a history of wage regulation.

Part Four, consisting of only one chapter, is an evaluation of the effectiveness of the different systems of wage regulation as anti-inflationary tools. Since wage regulation systems must function in a certain environment determined primarily by the management systems of the economy, the suitability of each system (centralised or decentralised) for coping with inflationary pressures is also examined.

From all this it follows that Part One deals primarily with macroeconomic problems, whereas the rest of the study is concerned with microeconomic issues, which are the real focus of the study.

A topic of this kind calls for a definition of inflation, which is of course a macroeconomic relationship. However, its causes cannot be explained solely by macroeconomic relations; economic activities in the microsphere ultimately determine what happens in the macrosphere. On the other hand, to understand properly the microsphere, it is first necessary to clarify corresponding macroeconomic relations.

Or, to put it in more concrete terms, our examination of wage regulation at the enterprise level, which the authorities also apply for the purpose of avoiding inflation, presupposes first a clarification of this term.

The countries under review do not follow an identical price policy. Some countries pursue a flexible policy; they use prices as an instrument for coping with real or potential market disequilibria. Those countries experience upwards creeping prices. On the other hand, some countries follow a rigid price policy; once consumer prices are fixed, they are not adjusted frequently enough to reflect the relationship between demand and supply. Those countries prefer to stick to stable prices even at the cost of queues at the shops, and other phenomena connected with shortages of consumer goods, rather than allow price increases. If prices cannot increase despite shortages, this is a case of repressed inflation. To be able to cover both kinds of inflation (open and repressed) with one definition, one needs a definition other than the traditional one which views inflation as a rise in the general level of prices. For this reason it is necessary to agree with A. I. Katsenelinboigen, who defines inflation as a '... process in which incompatibility arises between the paid-out sums of money and the value of goods available for sale'.[2] (Of course, it should be added that savings must be considered.) This means that the causes of disequilibrium and their manifestations in open or repressed inflation must be looked for in either side of the 'equation'.

There is no agreement among economists about the causes of inflation, as the international seminar held in Venice in 1974 strikingly showed.[3] For example, G. Garvy argued that 'four main types of causes of inflation can be distinguished in socialist countries: demand, cost, imported (a special case of cost) inflation and bottleneck inflation'.[4] By contrast P. Wiles contended that centrally planned economies have no wage-push inflation since there is no free collective bargaining and wages are supposed to lag behind productivity.[5] In his recent book he argues, as he had already indicated before, that the 'fundamental cause (of inflation) is something almost without parallel in market economies: the tolerated cost overrun, or the reluctance to bankrupt state enterprises'.[6] To him the most frequent reasons for cost overrun are shortages of skilled labour, stockpiling of goods due to their low quality and investment overstrain.[7]

Since this is not a study of the causes of inflation, we intend to express only briefly our views on this topic without attempting to go into any details. In our opinion over-investment is the most important factor of inflation. That it can play such an important role is the result of the Soviet-type strategy of growth, mainly in its two most important features: the drive for maximum economic growth and the great emphasis on the preferential growth of producer goods (in which military goods play an important role). This growth strategy often leads, as experience shows, to stepping up the investment ratio to a level which can be rightfully called over-investment. At that level barriers to growth are bound to occur (shortages of raw materials or capital goods or labour, or balance of payment troubles or disruptions in the coordination process, or a combination of some of these factors) and they necessarily result in bottlenecks which may be a source of inflationary pressures.[8] Over-investment also leads to an extension of the time needed for the completion of investment projects and thus to increases in the volume of unfinished investment projects (a very well-known disease of the Soviet-type economy).[9] Under conditions of a rapidly developing technology, the excessive length of completion time makes many investment projects partially obsolete before they can even be put into operation,[10] thus reducing returns.[11]

The delay in putting investment projects into operation can in itself produce inflationary pressures. This danger is compounded by the fact that such a delay particularly affects projects for production of consumer goods as a result of the preferential treatment of producer goods industries. If the fulfilment of investment plans is endangered for whatever reason, the planners will usually see to it that producer goods industries get preference in the allocation of machines, raw materials and labour. In practice, this usually means that wage payments are not matched with a proper flow of consumer goods.

The strategy of growth, as discussed briefly above, is the most important reason for market disequilibrium, but not the only one. Market disequilibrium in Soviet-type economies is often due, not to a disequilibrium in aggregates, (i.e., between purchasing power and a supply of goods and services) but to a disequilibrium in the composition of demand and supply in time and space. Shortages of consumer goods usually coexist with huge inventories of unsaleable goods.[12] The unsatisfactory product mix of consumer goods is primarily a result of deficient planning and shortcomings in the incentive system. It may be simply a consequence of gross errors in the forecast of changes in demand, combined with a disrespect for the established principle that the unpredictability which exists in the domain of demand changes must be cushioned by some reserves.[13] Regional disparities in the distribution of consumer goods,[14] caused by bureaucratic incompetence or other reasons, have also to share some of the blame for market disequilibria. It is known that the incentive system often encourages enterprises to a product mix which is contrary to demand. The inflexibility in prices in the face of changes in demand makes the situation even worse.

If we add up the two factors, over-investment and insufficient supply (the second being to a great degree the result of the first), it can be argued that the main causation of inflationary pressures runs from a failure to produce consumer goods and services to an extent which would match increases in wages.[15]

We have concentrated only on the main causes of inflation. In our study we will touch on some other causes (such as, e.g. the distorted price system and increases in prices on foreign markets) and we will, of course, discuss the role of wages. Despite this, the causes of inflation are not exhausted; this is, after all, not the objective of this book.

In conclusion to the introduction, some definitions of terms used in this study will be clarified. Many economists use the terms wage-fund and wage-bill as synonyms. We reserve the term 'wage-bill' for enterprises or associations and the term 'wage-fund' for larger aggregate units, mainly for the economy as a whole. For the purpose of our study it would be correct to use the comprehensive wage-fund which includes all wages and bonuses paid out to all employees (including the army and the security branches). The USSR and other countries use this concept of wage-fund for planning the income side of the balance of personal money incomes and expenditures, a balance which has to ensure market equilibrium.[16] Therefore, as long as we discuss wage fund on a theoretical level in connection with inflation, we naturally have in mind the comprehensive fund. However, the wage-fund which is used in Soviet bloc statistics for the computation of average wages for the economy is smaller than the comprehensive fund since payments of some employed groups are not included (military service personnel, security branches, party apparatus, etc.).[17] In this study average wage figures for the economy (or industry) are taken from official statistics or from other sources which have also used official statistics.

The enterprise wage-bill is the sum of paid out wages and bonuses during a certain period. The reforms of the second half of the 1960s brought about the emergence of the bonus fund as a separate fund (in Poland only for the managerial staff) from the wage-bill, and therefore it is necessary to distinguish the basic wage-bill and the bonus fund. The basic wage-bill corresponds to the sum of paid out basic wages (which include wage rates and basic salaries), payments for overfulfilment of performance norms, additional payments to time rates, bonuses for individuals as long as they are not paid out from the bonus fund, and finally payments for overtime and difficult working conditions as well as the equivalent of payments in kind.[18] In this study the wage-bill always covers a period of one year. If not otherwise indicated, we always have in mind the basic wage-bill when talking about the wage-bill in connection with the reformed systems of management and/or of wage regulation.

At the enterprise level average wages, basic average wages and wage rates should be distinguished. Both average wages and basic average wages have the same denominator—average number of registered (employed) staff, but they differ, of course, in the calculation of the numerator. In the first case the numerator is the basic wage-bill plus bonus fund, and in the second only the basic wage-bill. In order to avoid any misunderstanding in this study, whenever the object of discussion is average wages at the enterprise level during and after the reforms of the second half of the 1960s, the terms 'employment incomes' or 'total average wages' are used. The term 'average wages'—if not otherwise indicated—stands for basic average wages. Wage rates are explained in Chapter 3.

The regulation of wages and bonuses is enforced with the help of success (evaluation) indicators. They fulfil, however, different functions depending on the system of management and wage regulation, which will be discussed primarily in Chapters 3 and 5. Here, only briefly: generally it could be argued that in a centralised system where growth of wages and bonuses depends on the fulfilment of the plan targets, indicators serve as yardsticks both for assigning targets and for later comparing actual performance with the targets. In a decentralised system where binding targets are not assigned, they serve simply for measuring performance. In both systems performance as measured by these indicators is one of the main factors relevant to the growth of wages and bonuses. This circumstance is supposed to make them both a control over growth of wages and bonuses, and a stimulus to a greater effort.[19]

The countries under review use the same as well as different indicators. As will be shown, the choice of indicators is also influenced by systemic differences. There is no longer a uniform grouping of indicators. The traditional Soviet classification usually distinguishes two groups: quantitative and qualitative.[20] The first group includes all the indicators which characterise quantitative aspects of performance.

Gross value of output, commodity production and sales are typical quantitative indicators. The second group encompasses indicators which characterize fully or only partially the efficiency of performance.[21] Polish economists classify indicators into two groups: synthetic and special. J. Zielinski defines synthetic indicators as indicators which take both cost and revenue into account.[22] (Profit is the most synthetic indicator.) All other indicators are termed special. The mentioned classifications are not identical. In the Soviet classification, for example, labour productivity is a qualitative indicator, whereas in the Polish it is not. In our study we will use all the classification terms since this allows the nature of indicators to be best characterised.

Finally, it should be mentioned that in this book—for reasons which will be given later on—the term 'national income' is used in its Marxist concept, as it is understood in the countries examined.

Part One

1 Price Stability Policies in the Soviet Bloc Countries

In this chapter we will briefly survey the development of consumer prices in the countries under review and, above all, the policy followed towards price stability there since the start of economic planning. The period involved is marked by three phases which could not occur simultaneously in all the countries due to the earlier beginning of planning in the USSR. The first phase covers the initial phase of 'socialist' industrialisation; the second phase starts with the stabilisation of prices and ends with the abandonment of the price-cutting policy; the third phase lasts up to the present. In the initial phase of 'socialist' industrialisation all the countries (except East Germany) suffered from severe inflation. In the Soviet Union this occurred in the 1930s, during the three five-year plans, and continued during and after the Second World War up to 1947, while in the other countries it took place in the period 1949–50 to 1953.

All the East European countries applied to a greater or lesser degree the Soviet methods for achieving price stability and an elimination of the rationing system. First, they supplemented the consumer goods rationing system—a heritage of the war—by a free market where many goods were distributed at prices which were substantially higher than the rationed prices. It was hoped that, as the supply of consumer goods improved, it would be possible to reduce the differences between the dual prices and eventually to embark on a distribution of goods by the market. These hopes were dashed by the substantial increases in output targets in 1950 which aggravated market disequilibria. As in the Soviet Union the elimination of the rationing system and price stabilisation were achieved by a currency reform which confiscated a great part of the purchasing power of the population.

The currency reforms were followed by a policy of planned price cuts of consumer goods. The USSR embarked on this policy in 1947 and other countries in 1952–3. (East Germany had done so earlier.) The 1956 events in Hungary and Poland brought the uniform price policy to an end. Both countries abandoned the price cutting policy and took the first steps towards more flexible pricing. In practice this meant that in some periods prices crept upwards only very moderately, while in other periods they advanced more rapidly. After 1968 Hungary in particular embarked on a so-called mixed price policy; that is, it has retained some important elements of the rigid Soviet price system, but has complemented them with some elements characteristic of the market economy. After abandoning the policy of price cutting in 1954, the USSR has pursued a policy of rigid price stability. East Germany has followed the Soviet pattern, and Czechoslovakia joined them after the economic reform aborted due to the Soviet-led invasion.

In the following pages we will discuss the individual phases, starting out with a brief explanation of the Soviet-type price system as it was applied in all the countries of the Soviet bloc up to the 1960s and is still applied in the USSR, GDR and Czechoslovakia. Deviations from the Soviet-type price system will be explained later.

SOVIET-TYPE PRICE SYSTEM

Prices in a Soviet-type economy are fixed and can (with some exceptions) only be changed by the central authorities. They are not supposed to react swiftly to changes in demand and supply. Unlike in a market economy, their function is not to allocate resources, something which is done by state plans, though changes in prices do have some allocative effect.[1]

Producer and consumer goods prices are fixed differently. Prices of the latter, which are of primary interest to us here, contain turnover tax, whereas the former do so only exceptionally. The turnover tax rates are much differentiated[2] and were in the past very high on the average. For example, in 1956 in Czechoslovakia the turnover tax made up 48.3 per cent of consumer prices on the average.[3] In the meantime, the share of the turnover tax was reduced in all the countries by various reforms of wholesale prices and adjustments in consumer prices. The turnover tax is no longer, as it was in the 1950s, the main source of government revenues.

Apart from being a source of government revenues, the turnover tax is intended to fulfil several functions. It is usually argued that the tax rate differentiation is supposed to help equate supply of and demand for consumer goods. No doubt the original size of the tax for different consumer goods was set with this consideration in mind. Since consumer prices are inflexible and can be changed only with the explicit approval of the authorities, this function is considerably impaired.

Another function of the turnover tax is to dampen the level of the purchasing power of the population. For reasons of motivation, wages have been fixed higher and often increased by greater rates than governments deemed warranted from the point of view of the distribution of income. The difference has been recovered through prices of which the turnover tax is an important component.

What is more important for our purpose is the way the turnover tax is constructed. It is mostly the difference between retail prices (reduced by the commercial margin) and the wholesale prices. This method of construction means that the two circuits are separated and have a relatively autonomous development. This enables the government to change wholesale prices without affecting retail prices.[4] What is really changing in such cases is the turnover tax. In the case of wholesale price increases, the turnover tax diminishes; the same happens if retail prices are cut, and vice versa. If the wholesale price exceeds the retail price, the government uses subsidies, a very frequent measure. By breaking the linkage between wholesale prices (costs) and retail prices, governments have a powerful tool with which to shield the economy from possible inflationary pressures. The economy is also protected against imported inflation by the insulation of domestic prices from the effects of foreign market prices. Needless to say, such control of prices has also its negative consequences.

An integral part of the price system is the way the turnover tax is distributed among prices. In the Soviet Union low tax rates were fixed on food, with high rates on industrial goods, mainly clothing and footwear (but not on books and other items serving educational as well as propaganda purposes), and very high rates on alcohol and tobacco. Within individual groups of goods there are great differences in rates; generally, luxury and high quality goods are taxed more heavily than low and medium cost goods. With some exceptions no turnover tax is levied on services. The Soviet policy of high taxes on clothing has its origins in light industry's low capacity, which is the result of the applied concept of industrialisation. But it was also due to a desire on the part of the government to have at its disposal tools for taxing the agricultural population.

The Soviet rules for distributing the turnover tax among prices of consumer goods have been taken over in substance by the countries of the bloc.[5] The distorted price system which resulted from this distribution is—as will be shown later—one of the factors which pushes the economy in the direction of market disequilibria.

INFLATION IN THE INITIAL PERIOD OF ECONOMIC PLANNING

As has already been mentioned, all the countries apart from the GDR were plagued by inflation in the initial period of planned economy. The USSR experienced the most severe and protracted inflation; in the period 1928–40 retail prices in state and cooperative stores increased twelvefold.[6] In Czechoslovakia, Poland and Hungary, inflation was not nearly so rampant.

TABLE 1.1 Evolution of retail prices (annual rates of increases in per cent)

	1932–7	1938–40	1945–7	1949–53
USSR	33.5	8.5	43.5	
Czechoslovakia				14.0[a]
Hungary				22.2[b]
Poland				18.5[c]

[a] Refers to 1948–53.
[b] Refers to 1949–52. The single huge price increase in 1951 is included.
[c] The single huge price increase in 1953 is included.

SOURCES
computed on the basis:
USSR: G. Garvy, *CESES* pp. 316–17 (Kolkhoz market prices are not included).
Czechoslovakia: Č. Kožušník, 1964, p. 96.
Hungary: *Statisticheskii sbornik*, 1959, p. 121.
Poland: *Rocznik statystyczny*, 1957, p. 238.

In all the countries reviewed, inflation, it seems, was not a result of conscious policy. There is no evidence that inflation was used as a deliberate tool for financing industrialisation. This is more true of the USSR than of the other three countries. It is known that all the East European countries imitated the Soviet model in every respect. Why not assume then that these countries also imitated the Soviet Union in this respect even if, in the USSR, inflation was not an intentional product of industrialisation?[7]

It seems, however, that the countries did not make serious efforts to arrest inflation.[8] The effort of the planners was primarily directed towards retaining a distribution of national income which would ensure rapid industrialisation; price stability was regarded as a secondary matter.[9] Apparently, inflation was, in a sense, welcome; it enabled income to be redistributed in favour of the financing of industrialisation.

The ambitious five-year plans with their great stress on heavy industry were the main cause of inflation in the Soviet Union (and also in East European countries, as will be shown in Chapter 2). The implementation of these plans assumed increases in productivity and in the availability of machines and materials to an extent which soon turned out to be unrealistic. The results of the first Soviet annual plan (1928–9) were in many respects below expectations; the targets for consumption in particular were not fulfiled. Though some economists advocated moderation, the Soviet leaders chose the opposite direction. They decided to increase the output targets of the five-year plan and to speed up collectivisation.[10] The new target for 1931 was fixed at a level nearly double that of the original plan.[11]

The acceleration of industrialisation necessarily meant an increase in the investment ratio. According to N. Jasny, investment in the socialised sector increased close to fourfold at current prices during 1928–32. (At the same time investment costs increased by 25 per cent).[12] Such a rapid increase in the volume of investment was inevitably at the expense of consumption, all the more because increases in investment were primarily channeled into heavy industry. According to M. Dobb, increased investment in heavy industry was achieved not only at the expense of investment projects planned for light industry but also, in some cases, at the expense of the current output of light industry.[13]

It has already been mentioned that productivity targets were unrealistic.[14] In order to achieve output targets, employment was increased by 97 per cent in the first five-year plan period, almost two and a half times as much as was envisaged in the original plan.[15] With such a rapid increase in employment, an over-expenditure of the wage fund could not be avoided. Due to the great demand for labour and to price increases, average wages also grew faster than anticipated in the plan. The demand for consumer goods increased rapidly; however, consumer goods industries did not expand correspondingly. The five-year plan target for consumer goods was under-fulfilled,[16] one of the main reasons being the large-scale slaughter of livestock by the peasantry as a response to forced collectivisation. It is obvious that under such conditions inflation could not be avoided. The situation in the period that followed till 1940 was not different in principle.

A glance at Soviet statistical figures reveals, as could be expected, that the acceleration of industrialisation also accelerated the rate of inflation. In Czechoslovakia, Poland and Hungary the increase in output targets in 1950—a result of the intensification of the cold war—brought the short-lived and fragile market stability to an end. In 1951 Hungary and Poland were forced to reintroduce and Czechoslovakia to expand the rationing system and to increase prices, which brought a new inflationary wave into being.

PRICE STABILISATION POLICIES

After the Second World War all the countries suffered from shortages of consumer goods, a condition which fueled inflation. With the adoption of the Soviet concept of economic development, the situation in the East European countries became even worse. No wonder the black market flourished. Authorities were confronted with a great challenge: how to ensure a proper distribution of the scarce consumer goods, a distribution which would at the same time pave the way for a renewal of market equilibrium.

As early as the 1930s the Soviet Union used a dual price system for this purpose.[17] The rationing system, based on low prices, was supplemented by a commercial market where goods were available at much higher prices (commercial prices). East European countries took over this system as they did other tools of economic policy. Commercial prices had to fulfil several functions:

1. To compete with and gradually eliminate the black market. What was no less important, they were supposed to be instrumental in getting hold of a large part of the profits—for government coffers—which the black-marketeers were making.

2. To draw off excess purchasing power[18] and thus pave the way for the abolition of the rationing system and the renewal of the distribu-

tion of consumer goods through the market. This was supposed to be achieved by a gradual closing of the gap between the two levels of prices, mainly by reducing commercial prices.

3. Commercial prices became a new instrument for taxing the population and thus a new source of accumulation for the then-beginning planned economy.

4. They also had a very important institutional function: to tax heavily the self-employed population, mainly the capitalists, and in such a way reduce their economic power. Rations were differentiated according to different criteria, an important one being class background. The rationing system discriminated against the self-employed. In Czechoslovakia approximately 20 per cent of the population was, at one time, excluded from the rationing system and had to cover its needs on the much more expensive commercial market.[19]

As early as 1948 East Germany introduced the commercial market where the initial prices were approximately at the level of the high black market prices.[20] This enabled her to embark on a policy of price cutting, in some years carried out several times for different groups of consumer goods, always accompanied by a great propaganda barrage. Therefore, in East Germany—unlike Czechoslovakia, Hungary, and Poland—the implementation of the first five-year plan was accompanied by price reductions rather than by price increases. To what extent the announced price reductions meant a genuine increase in purchasing power is difficult to judge.

It is known that Czechoslovakia wanted to take the same path as East Germany, but failed. It too started with reductions of the commercial prices introduced in 1949, but after a huge increase in 1950 in the five-year plan's output targets, mainly for heavy industry, commercial prices again started to increase.[21] Similarly Poland failed; even a strict money reform effected in 1950 could not bring about a stabilisation of prices, except for a very short period.

Hungary was the only country that applied the dual price system for only a short period (from March to December 1951).[22] The hyperinflation which Hungary experienced and which in a short time depreciated the purchasing power of the *pengö* to a negligible fraction radically solved some of the problems which even the dual price system and currency reforms had not been able to achieve. It virtually wiped out all cash holdings and bank deposits and thus created favourable conditions for a renewal of market equilibrium. In addition, at the time of the currency reform of 1946 which put an end to the hyperinflation, new wages compared to new prices were set at a level far below the prewar one.[23] But very soon inflationary pressures started to develop again. In December 1951, in connection with the abolition of the rationing system, the Hungarian government put through a second quasi-currency reform without changing the bank notes. Prices of consumer goods were increased by 50–100 per cent, whereas wages rose by only 20 per cent.[24] Even this drastic reduction in real wages did not bring about price stability.

A turning point in the price stabilisation policy came in 1953. Czechoslovakia, Poland, and Hungary managed to bring price increases to a halt and embarked on a policy of price cutting following the Soviet pattern. Several circumstances contributed to this new development. Rapid industrialisation without due consideration to the material interests of the population (reflected in a drop in real wages which was partly mitigated by a rapid growth in employment) generated political tension. This was enhanced by the increasing imposition of strict controls over all facets of political life, which were enforced by methods bordering on terror. Fear of unrest induced governments to look for ways to improve the material condition of the population. Stalin's death, together with the changes in Soviet foreign policy and in economic priorities (greater stress on consumer goods industries) enabled East European countries to slow down the pace of industrialisation, something which was needed in any case in order to cope with imbalances cumulated during the first plan period. They allowed the investment ratio to be reduced and, with it, the rate of growth of employment, and released resources for consumption. Also of importance was the fact that in 1953 the introduction of the Soviet price, wage and tax system with its strict controls from the centre was more or less completed or brought to completion by the price reforms. In the meantime, the planners learned on the job how to plan, manage and control a Soviet-type economy.

In Czechoslovakia this new policy was preceded by a package of reforms—aimed at a renewal of market equilibrium—known as a currency reform. It included, apart from a currency reform, a price reform, wage adjustments and abolition of the rationing of consumer goods. The dual price system was replaced by uniform prices which on the whole were higher than the weighted average of both.[25]

In Poland the rationing system and the dual price system were abolished much earlier, in 1949, but were reintroduced in 1951 after the currency reform of 1950 failed due to an acceleration in economic growth. In 1953 Poland carried out a price reform which was, in a sense, a currency reform without changing bank notes. Prices were increased by 40 per cent while wages rose only by 30 per cent. At the same time rationing of consumer goods was again abolished, this time permanently.[26]

East Germany joined the new policy in the People's democracies by systematising its price reductions.[27] Unlike other countries, the abolition of the rationing system was delayed until 1958 and was preceded by a currency reform in 1957.

Currency Reforms

Before the policy of price reductions is examined, currency reforms will be discussed. They played a very significant role in the effort to renew market equilibrium, which was undermined first by war events and later by ambitious national plans. In this study we will confine ourselves to the discussion of currency reforms which came into being due to the second factor,[28] namely the Polish reform of 1950, the Czechoslovak of 1953, and the East German of 1957. Exceptionally, we will also discuss the Soviet currency reform of 1947, which was a result of war events.

The currency reforms in Poland, Czechoslovakia, and East Germany were more or less shaped according to the principles of the Soviet currency reform of 1947, though reflecting in their implementation special conditions prevailing in individual countries. Not only this, Soviet experts probably had an important role in their detailed conceptualisation. It is positively known that the Czechoslovak currency reform was carried out under the guidance of Soviet experts. All the currency reforms followed two goals in substance:

1. They aimed at stopping inflationary pressures by removing the currency overhang and at a renewal of the distribution of goods through the market. Rationing of goods combined with a commercial market is not an ideal institution for a 'socialist' state. The administration of rationing is expensive, but what is worse, it hinders the promotion of economic efficiency by undercutting incentives to higher productivity. The money equivalents of rationed goods—due to their relative cheapness—can be earned without great effort. The higher the rations and the bigger the difference between rationed and commercial prices, the smaller the incentive to work hard since increments in earnings can be turned into more consumer goods only on the very expensive commercial market.

The question can be raised as to why these countries resorted to currency reforms. They could also have achieved the renewal of market equilibrium by price adjustments. Such a solution would have been politically unacceptable; it would have caused great hardship to the population, primarily to workers and low-income groups that are the social base of the system. It would also be contrary to the second goal of the reforms. In addition the currency overhang was so big in some of the countries that the new prices would have had to be on the level of, or even above the level of, commercial prices. According to E. Ames, the governments in such countries would have been hesitant for ideological reasons to rely solely on price adjustments.[29]

2. They were also to contribute to a strengthening of 'socialist' ownership relations by weakening the economic power of the remnants of the bourgeoisie and the petty bourgeoisie.[30] For this purpose the currency reforms were designed in such a way that the burden of the reform would be shifted as much as possible to the shoulders of the owning classes, including speculators.[31] In the USSR, where the private sector no longer existed, the reform was intended to affect speculators as well as those collective farmers who had managed to make money on the lucrative collective farms market.

In order to achieve these goals the reforms confiscated a great portion of the money holdings of the population. The extent of confiscation and the method of its implementation varied from country to country. In the USSR, Poland and Czechoslovakia, cash holdings in particular were converted at a much more unfavourable rate than wages and prices. In the USSR wages, pensions and procurement prices for agricultural products remained at the pre-reform level. Unified prices for consumer goods were fixed in different ways: new prices for food were mostly set at the level of rationed prices (in the case of bread and some other foods prices were even reduced); however, industrial goods were fixed at a level apparently closer to high commercial prices. Cash holdings were converted at the rate of 10 (old roubles) to 1 (new), whereas savings deposits up to r3000 at 1 : 1, and deposits above this sum at a more unfavourable rate.[32]

In Poland wages, prices, rents and claims of the state sector were converted at a rate of 100 : 3, but the conversion rate for cash holdings and claims of the private sector was 100 : 1. Savings deposits up to Zł100,000 were converted at the rate of 100 : 3, deposits above this sum at a rate of 100 : 1.[33]

In Czechoslovakia wages, pensions, and claims of the state sector were converted at a rate of 5 : 1. Unified prices were increased by 15 per cent of the weighted pre-reform average divided by five; prices of foodstuffs were increased by 29 per cent while industrial goods remained practically at the old level (divided, of course, by five). Cash holdings up to 300 korunas per person were converted at a rate of 5 : 1, sums above the limits of Kčs300 and claims vis-à-vis the state sector at a rate of 50 : 1. On the other hand, savings deposits up to Kčs5000 were converted at a rate of 5 : 1, deposits above this sum at a graduated ratio.[34]

In East Germany wages and prices did not change; the unification of prices and the abolition of the rationing system were not achieved until 1958. Cash holdings up to DM300 per person were converted immediately, the remainder only if there was no suspicion that the money was derived from speculation.[35]

The second goal was achieved by various methods:

1. By treating large cash holdings differently from small. This was the case in Czecholsovakia.

2. By relying on the fact that the provision allowing much more favourable conversion rates for wages and prices than for cash holdings would affect more the self-employed and the speculators, who

TABLE 1.2 Currency reforms (new currency units per 100 units of old)

	USSR Dec. 1947	Poland Oct. 1950	Czechoslovakia June 1953	GDR Oct. 1957
Cash holdings				
lower portion	10	1	20	100
upper portion	10	1	2	100
Saving deposits				
lower portion	100	3	20	100
upper portion	sliding scale	1	3.3	100
Wages	100	3	20	100
Prices		3		100[a]
food	at the level of rationed		29% higher	
industrial goods	69% lower than commercial		at the pre-reform level	
Claims of the private sector		1	2	

[a] The elimination of rationing and two level prices was not carried out until 1958. At that time prices for rationed food were fixed somewhere below commercial prices. For this price increase compensation was given in the form of lower prices for some industrial goods and through an increase in wages. See *Der Arbeitergeber*, 20 June 1958.

SOURCES
See Notes 32–5 and N. Spulber, 1957, p. 128.

owned more of the cash holdings. This was the case in Poland[36] and Czechoslovakia as well as in the USSR.

3. By fixing a higher conversion rate for government or government institution claims (taxes, bank loans) than for the claims of the private sector (Poland and Czechoslovakia).

4. By granting a much better conversion rate for savings deposits than for cash holdings. It was assumed that the self-employed and the speculators do not entrust most of their holdings to the banks for fear that they might be questioned about the origin of the money[37] (see also Table 1.2).

The more favourable treatment of savings also served another purpose. It aimed at encouraging consumers to deposit their savings with banks as a way of reducing pressure on the consumer market.

With the exception of East Germany, the currency reforms also affected many blue- and white-collar workers by depriving them of the core of their savings. Blue-collar workers were, in addition, hit harder by the abolition of the rationing system, mainly those who were engaged in heavy industry and who got privileged rations for low prices. In Czechoslovakia the new price relations between foodstuffs and industrial goods favoured the peasantry at the expense of the employed. In the USSR the reverse was the case.

Policy of Price Cutting

Price reductions, first introduced in the Soviet Union and later imitated by other countries, were the result of conscious government policy. During the currency reform, prices for many goods were fixed much above what was assumed to be the equilibrium price in order that they might be later reduced.[38] In the USSR the abolition of rationing prices was preceded by huge price increases.[39] In addition great stocks of consumer goods were accumulated, partly to ensure the success of the currency reform and partly to prepare for the ensuing price reductions. By lowering prices at approximately the same time of year in most of the countries, the governments tried to make the price cuts appear to be annual government 'dividends' from annual performance. Needless to say, these price cuts were given great publicity and were used as a device to prove the superiority of the socialist system over the capitalist system. Some writers even went so far as to hail the price cuts as a socialist way of increasing the standard of living.

Priority seems to have been given to price cuts rather than nominal wage increases for a variety of reasons—political, economic and ideological. It is not so clear what the political motive was for this move in the USSR. Probably propaganda considerations, including the desire to impress workers abroad, played an important role. As to the motives of other countries, these are much clearer. Apart from the fact that this policy was applied at a time when all the countries were trying hard not to deviate from the Soviet model, the main political motive was that this method of increasing the standard of living might back up the regime's authority which had been shaken by the results of the first medium-term plan with regard to consumption. Among economic reasons was the consideration that price cutting might help move consumer demand in the direction of the planned consumption structure. This consideration was all the more important in that governments were interested in getting rid of the aforementioned accumulated stocks of goods. It was also believed that a deflationary policy would be an effective barrier to the inflationary pressures which had plagued the economies of the area up to that time.[40]

Price reductions varied from country to country in extent and in length of time of operation. In the USSR this policy was applied for several years (1947–54) and included seven price cuts, which reduced prices by more than 50 per cent.[41] In Czechoslovakia and the GDR price cuts lasted until 1960; in the former they accounted for a decline of 14.8 per cent in the cost of living if the price reduction of 1953 is disregarded, and in the latter 46.3 per cent if 1950 is taken as the base year. In both countries the pace of price reductions slowed down remarkably after 1955. In Poland[42] and Hungary the policy of price cutting was short-lived and was of smaller importance (see Table 1.3).[43]

The price reductions did not contribute to market stability as had been hoped; on the contrary, they became a factor of disequilibrium.[44] The selection of goods for price decreases was not always governed by market considerations; political aspects played an important role. Often goods were reduced in price when the economy was not able to ensure a sufficient supply of them even at existing prices. In addition, the policy of price reductions slowed down the growth of money wages[45] and became an obstacle to the utilisation of wages as an incentive. Wage increases can be differentiated by sectors, branches, enterprises and even occupations, according to the need for stimulation, whereas price reductions benefit everyone who buys the goods which have been lowered in price.

In some of the countries, the slow growth in money wages hampered efforts to correct wage differentials, which had narrowed excessively during the first five-year plan. Again, in other countries it hindered plans to raise agricultural procurement prices and low wages.[46,47]

TABLE 1.3 Price reductions

	1947	1948	1949	1950	1951	1952	1953	1954	1955	1956	1957	1958	1959	1960
USSR	100	86.7	77.4	64.1	59.5	57.2	52.0	49.1	58.1	57.4	56.9	54.4	53.3	52.7
GDR				100	78.2	67.9	64.2	60.9	94.1	91.5	89.6	89.5	87.2	85.2
Czechoslovakia							100	96.7	94.3	93.3				
Hungary							100	95.1	91.6	90.6				
Poland							100	93.8						

Indexes

SOURCES
USSR: G. Garvy, *CESES*, pp. 316–17 (refers to the retail price index, kolkhoz market prices are not included).
GDR: *Statistisches Jahrbuch der DDR, 1976*, p. 29 (refers to the retail price index for the whole population).
Czechoslovakia: *Statistická ročenka ČSSR, 1976*, p. 41 (refers to the cost of living of workers' and employees' households).
Hungary: *Statistical Yearbook of Hungary, 1967*, p. 245 (the same as in Czechoslovakia).
Poland: *Rocznik statystyczny, 1957*, p. 238 (refers to prices of goods and services acquired by the population).

DIVERSE PRICE STABILITY POLICIES

Soviet and GDR Price Policy

After 1954 the USSR, as already mentioned, embarked on a policy of rigid price stability.[48] By 'rigid' is meant that the general level of prices is not increased even at the cost of market disequilibria. Changes in prices can be made only by the price authorities, and they—as experience shows—are reluctant to increase prices unless the discrepancies between demand and supply are considerable. This policy is ideologically supported by the thesis that under socialism prices should move only downwards in the long run.

This does not mean that no price adjustments have been made. In the USSR some have occurred almost every year in the period since 1954. Downward price adjustments serve much the same purpose as bargain sales in the West. Briefly, their object is to clear the inventories of old-fashioned and low quality goods which consumers are reluctant to buy at current prices, or to speed up the sale of goods which are going out of production and are to be replaced. Some price lowering is due to a considerable decline in the production costs of the goods and the expectation that such a price change would considerably increase demand. Such an approach is taken to goods whose production can be easily expanded, thus contributing to an easing of pressure on the market.

Price reductions are also put into effect to make price increases more acceptable. It has already become a custom in the East to sweeten the bad news of price increases with some price reductions. For example, the huge Soviet price increases for meat and butter in 1962 were combined with simultaneous price decreases in sugar and textiles.[49] More recent price increases (January 1977) have also been coupled with some price decreases.[50]

In the Soviet Union and the GDR for a long time, and in Czechoslovakia in the seventies, the cost of living has shown no change, as Table 1.4 reveals. This does not mean that their economies work in a way that does not generate inflationary pressures. The Soviet-type economy manages fairly well to shield the economy from wage-price and price-wage spirals. It has at its disposal tools to prevent possibly excessive wage increases developing into price increases, and vice versa, which would set an inflationary spiral into motion. However, despite much effort the Soviet-type economy is marked by widespread shortages in some goods and services, which coexist with surpluses in

TABLE 1.4 Cost of living (annual rates of increases in per cent)

	1951–5[a]	1956–60	1961–5	1966–70	1971–5
Czechoslovakia	5.6	−1.8	0.6	1.7	−0.05
GDR		−1.7	0.02	−0.04	−0.1
Hungary	10.9	0.2	0.5	1.0	2.8
Poland	11.6	2.9	2.2	1.6	2.4
USSR	−5.2	0.2	0.2	−0.2	−0.04

[a] Figures for Czechoslovakia, Hungary and Poland for 1951–3 are taken from Table 1.1.

SOURCES
computed on the basis:
Czechoslovakia: *Statistická ročenka ČSSR*, *1975*, p. 41 (refers to workers' and employees' households).
GDR: *Statistisches Jahrbuch der DDR*, *1970*, p. 352 and *1976*, p. 29 (refers to workers' and employees' households).
Hungary: *Statistical Yearbook of Hungary*, *1967*, p. 245, *Magyar statisztikai zsebkönyv*, *1977*, p. 145 (refers to workers' and employees' households).
Poland: *Rocznik statystyczny*, *1957*, p. 238 (for 1951–5) (refers to prices of goods and services acquired by the population) and *1975*, p. XXXVI, (refers to the cost of living of an average working family).
USSR: Up to 1970 from R. Clarke, *Soviet Economic Facts 1917–70*, 1972, p. 14. Figures for 1971–5 *Narodnoe khoziaistvo SSSR*, *1975*, p. 643. (All figures refer only to price indexes for consumer goods bought in state stores.)

other goods. The shortages are not always in the same groups of goods, and they also vary in severity from period to period. Since prices are centrally regulated, shortages can push up prices legally only if the authorities so wish. The authorities very often prefer shortages to price increases. Therefore, a Soviet-type economy is exposed to repressed inflation which manifests itself in various ways, such as in queues in front of stores, in a black market for some goods and in paying prices for some goods which are higher than the official ones. Other indications of repressed inflation in the USSR are to be found in the higher prices on the collective farm market. M. Bornstein correctly alludes to several circumstances which one should keep in mind when drawing conclusions from this fact. One is that only some food items are sold on the collective farm market. Secondly, part of the higher prices is due to the higher quality and the freshness of the food available at that market.[51]

Usually, increases in savings are also mentioned as a sign of repressed inflation. In our opinion what may be indicative of repressed inflation is not simply a growth in savings at the same rate but an increase at a higher rate. But even higher rates can be partly due to new stimuli for voluntary savings. The fact that it is impossible to quantify the share of involuntary savings gives rise to various views which are more a reflection of *Weltanschauung* than of hard facts.

The official Soviet price index, since it does not take into account repressed inflation, does not give an accurate picture of price movements. In addition, as many authors note, it lacks accuracy for another reason too as the cost of living index in most of the countries of the Soviet bloc does. It is based on official prices[52] and does not take sufficiently into consideration what is often called increases in average prices in the East European literature, i.e., price increases which result from genuine or non-genuine product quality changes, from the introduction of new products, in brief, from shifts in product mix to more expensive goods.[53] Finally, the USSR—unlike the other four countries under review—does not publish a cost of living index, but only a price index for consumer goods which is usually less accurate in reflecting price changes relevant to the standard of living.

It is rather difficult to estimate correctly how much distortion there is in the Soviet price index due to repressed inflation and to inaccuracy in its design as an indicator of the cost of living. Some of the figures advanced are doubtless exaggerated;[54] to accept them would be to deny to a great degree that the USSR has made remarkable progress in the past two decades in raising the standard of living despite widespread goods shortages.[55] Figures indicated by G. Schroeder and B. Severin[56] and also by D. Howard[57] seem to be more realistic.

Growing state price subsidies are an important factor of the accomplishments in price stability. They are applied primarily to meat and milk. According to N. T. Glushkov, the Chairman of the State Price Committee, the subsidies for meat and butter amounted in 1975 to 19 milliard roubles,[58] which meant 8.9 per cent of government expenditures. The relatively high subsidies are mainly a result of increases in procurement prices in 1965 and 1970.[59] In 1962 Khrushchev passed along a great part of the increases in procurement prices to retail prices, but Brezhnev is reluctant to do the same. Recently there has been a tendency to limit subsidies for industrial goods and rather to use the resources for wage increases.[60]

As far as is known, the Soviet planners have never seriously considered eliminating the distorted price system for consumer goods as other countries have tried to do (as will be shown later). Apparently the subsidies do not cause as much concern for Soviet planners as they do for planners in other countries. The distorted price system does not cause as much trouble in balance of payments as in Poland, and the USSR does not intend to follow Hungary with regard to flexibility in price formation.

The GDR follows in substance a similar price policy to that of the USSR. In the 1960s when the GDR, first among the countries of the Soviet bloc, embarked on an economic reform there were some indications that it might also want to change its price policy. It carried out a reform of prices of producer goods in three stages (1964-6), the aim of which was to bring prices more in line with costs and thus to eliminate at least a portion of the relatively high subsidies. Prices of raw materials in particular were increased.[61] The price reform, however, did not affect consumer prices[62] though the distortions in the German price system did not differ substantially from those in other countries (see Table 1.5). The German leadership stuck to a rigid price stability. This policy was again confirmed in 1971—maybe under the influence of the unrest in Poland—by the decision of the Politbureau not to allow price increases in the period 1971-5.[63] Of course, this does not mean that all prices are really stable. Like the USSR, East Germany experiences a repressed inflation, the extent of which it is difficult to estimate. Again like the USSR, increasing costs of food are subsidised; according to one East German source prices of food include subsidies of 27 per cent.[64]

TABLE 1.5 Retail price index in the GDR (factory price + commercial margin = 100)[a, b]

	Average index	Dispersion
Foodstuffs	116	98-533
Luxury articles (*Genussmittel*)[c]	196	100-800
Footwear	148	100-325
Textiles, clothing	134	71-318
Other industrial goods	108	100-315
Consumer goods in total	127	

[a] The author identifies the excess over 100 as 'centralised net income' which is nothing else but the turnover tax.
[b] Computed on the basis of planned turnover for 1957.
[c] In the Soviet bloc countries this term refers to such items as alcohol, coffee, cocoa, tea, citrus fruits, etc.

SOURCE
H. Langer, *Wirtschaftswissenschaft*, no. 5, 1958.

Price Policy in Other Countries

The events of 1956 brought to an end the policy of price cutting in Hungary and Poland. In both countries dissatisfaction with the regime, building up over the years, came into the open after Khrushchev's revelations of Stalin's crimes; in Hungary it culminated in an uprising, and in Poland it was dissolved by a change in the leadership. In a situation so full of political tension it was difficult to resist wage increases, all the more since they offered themselves as a proper instrument to placate workers. Under such conditions it was unthinkable to continue price cutting. One of the principles of the new economic reform which was promised by the new Polish leadership was to give

TABLE 1.6 Turnover tax and subsidies on consumer goods in 1969 in Hungary

	Turnover tax (+) subsidy (−) in per cent of consumer prices	Turnover tax	Subsidies
		in milliard forints	
I. Foodstuffs and luxury articles	+4.2[a]	8.6	7.0[a]
Basic foodstuffs,	−36.0[a]	0.2	6.7[a]
meat and meat products	−45.5[a]		3.0[a]
milk and dairy products	−59.0[a]		2.0[a]
fat	−23.0		0.4
cereals	−25.0		1.0
sugar	—	—	—
luxury articles	+48.0	7.1	
II. Eating out	−4.0	0.3	0.7
III. Clothing,	+26.6	6.0	0.4
yard goods	+37.6	0.9	
ready made clothes	+21.1	1.6	
footwear	+9.4	0.5	0.2
IV. Miscellaneous industrial goods,	+13.6	4.4	1.8
furniture	+4.1	0.1	0.1
fuel	−53.2	—	1.6
medicine	+54.0	1.5	0.1
V. Paid services	−46.4	0.4	5.4
Total	+4.6[a]	19.8	15.3[a]

[a] The price changes of 1971 are taken into consideration.

SOURCE
1. Vincze, 1971, pp. 37–8.

the market a greater role in the formation of prices, and price cutting was in contradiction with these intentions.

On the contrary, in 1957 both countries embarked on a policy of price increases to reduce the purchasing power which had been swollen as a result of wage-push enabled by the political events in 1956. Poland was more daring in this regard than Hungary. Gomulka, thanks to his initial popularity, could more easily manoeuvre than Kádár.[65] In 1959 in connection with the reform of wholesale prices the Hungarian government also carried out some retail price adjustments

TABLE 1.7 Taxation of consumer goods in Czechoslovakia

	\multicolumn{2}{c}{Percentage share of turnover tax in the retail price}	
	1956	1967
Fuel	6.9	−35.7
Potatoes	23.0	−18.6
Eggs	23.4	−9.4
Fats	29.4	5.4
Meat	31.4	9.3
Milk	32.6	12.0
Fruits and vegetables	35.2	12.3
Flour products	37.1	13.1
Building materials	37.5	14.2
Household equipment and appliances	37.6	30.9
Sanitary goods and drugstore products	42.7	31.4
Footwear	45.0	32.8
Clothing	45.2	33.3
Cultural facilities	47.4	39.6
Other foodstuffs[a]	53.1	43.8
Means of transport	54.6	46.3
Alcoholic beverages	65.3	47.7
Sugar	67.1	50.8
Tobacco, cigarettes cigars	80.2	65.7
Total	48.3	24.4

[a] The groups 'other foodstuffs' includes coffee, tea, chocolate, cocoa.

SOURCE
J. Adamíček, *Politická ekonomie*, no. 8, 1963.

in the Soviet style; some prices were increased and some reduced.[66]

Both countries, as well as Czechoslovakia (mainly in 1966–9), tried to pursue two goals in the ensuing price adjustments: (a) one long-term which was to change the irrational price relations which had become a permanent source of market disequilibria and inefficiency and (b) one short-term which was to solve current market imbalances.

It has already been mentioned that the turnover tax is distributed very unevenly among the prices of consumer goods so that prices deviate from costs in different directions and degrees, and also that services for the most part are not burdened by tax at all. Tables 1.6, 1.7 and 1.8 show that basic foodstuffs in Hungary and Poland[67] are subsidised or, as in Czechoslovakia, taxed at a low rate.[68] On the other hand, clothing and footwear are heavily taxed and are therefore relatively high in price. The tables, except for Hungary, do not give figures on subsidies for services. It is, however, generally known that prices of services, mainly charges for transportation and rents for housing, were set at a low level.

TABLE 1.8 Relationship of accumulation[a] to costs of production of some consumer goods in Poland (in per cent in 1963)

Sugar	34	Personal cars and bicycles	30
beer	approx. 50	woollen fabrics	118
meat and meat products	approx. 5	cotton fabrics	18
butter and other dairy products	approx. 8	silk fabrics	47
milk	loss	flax fabrics	loss
coal, gas, electricity for the population	loss	footwear	21
television sets	23–46	hosiery	108

[a] Accumulation includes profit but mainly turnover tax.

SOURCE
J. Struminski, *Gospodarka planowa*, no. 2, 1963.

One of the reasons for this policy—as noted—was that prices were supposed to perform a redistributive function between the agricultural and non-agricultural population. But they were also supposed to redistribute income in favour of low income groups per capita and thus to reduce the differences arising from employment incomes and the number of children.[69] By imposing small tax rates on basic foodstuffs and setting low rents, the government wanted to make sure that low income groups would be able to buy essential goods and shelter at relatively low prices. In the period when the new price relativities were fixed,[70] the overall standard of living—disregarding the differences between individual countries—was relatively low due to the war, the industrialisation drive and armament. The burden of industrialisation and armament could not be evenly distributed if the governments wanted to preserve their claim to be defenders of the low income groups.

The consequences of this policy were very soon felt. The low prices for basic foodstuffs have engendered a considerably higher consumption of these goods than would have been the case at the existing income level, with price relativities corresponding to costs. The average consumer facing the problem of how to spend the money left over after his essential needs have been met, and wondering whether to buy expensive textile products, ready-made clothes and footwear, or relatively cheaper, higher quality foodstuffs, often prefers the last of these alternatives. Yet these countries either do not have the capacity to satisfy the growing demand for certain foods from domestic production or else they must limit their export of some goods in order to satisfy domestic demand. At any rate this is detrimental to foreign trade balance.

There is another reason why the discrepancies between costs and prices push the economy to an irrational structure of consumption and also production. Low prices for many services, e.g. rents for housing, push demand above the level the countries can afford because of the high investment involved, and, on the other hand, high prices hamper increases in demand for the products which the country could more easily produce.[71]

This is not the whole story. At the time when price relativities were fixed, East European countries pursued—according to the Soviet pattern—a policy of low agricultural procurement prices. They relied heavily on the Stalinist method of coercion. In the course of time they have come to realize, at the cost of great losses to the economy, that better performance in agriculture can only be achieved with more incentives. Since productivity in agriculture does not grow fast enough, East European countries—as well as the USSR—have had to resort to relatively frequent increases in procurement prices. Consequently, the turnover tax in many foodstuffs has turned into a growing subsidy, a fact of great concern to the planners.

At first glance it would seem that the solution is very simple—namely, to increase prices of foodstuffs and decrease prices of industrial goods. Under conditions of a slowly growing standard of living this is a politically risky and dangerous solution as the Polish unrest in 1970 showed. Such a general price reform would adversely affect the low income groups and pensioners in whose budget expenditures on basic foodstuffs and housing play the most important role. It can be assumed that such a reform, which cannot be combined with a clearly sufficient compensation to all who might be affected, would probably generate great tension if not unrest. It should be borne in mind that the public in East European countries—in some of them more than in others—has a general distrust of price reforms and tends to view them as a device to lower the standard of living.

Such a general price reform can be successful only if there is sufficient idle productive capacity in light industry, especially in the textile and footwear branches (or reserves in foreign currencies) to satisfy the newly emerging composition of aggregate demand. None of the countries under review has the material prerequisites for a swift change in price relations. Therefore, they must prefer a gradual, long-run solution. For example the Hungarian reform of 1968 promised that distortions in price relations would be gradually eliminated over the next 10 to 15 years.[72]

The short-term goal of price adjustments (more precisely price increases) was implemented in combination with price reductions as already noted. After some time the Polish[73] and Hungarian[74] governments made an important shift in this policy; they put much greater stress on compensation in money terms (increases in wages and pensions) for price increases. A more pronounced creeping inflation has become an acceptable solution to the plaguing problems. It turned out to be difficult to coordinate the timing of price increases and decreases and to find always, whenever price increases were felt necessary, goods which from an economic viewpoint would be suitable for price reductions. A more important motive for a change in policy may have been the governments' realisation that this combination was much discredited by their own propaganda.

The short-term and long-term goals of price adjustments, despite efforts, cannot always be in harmony. Short-term interests often push in the opposite direction to long-term. This is also the reason why Hungary—and the same is true of Poland and Czecholsovakia—has not been able to make discernible progress in its effort to change price relations; in fact quite the opposite happened.

Price policy has been complicated in recent years by explosive price increases in world markets, mainly for oil. True, not all the countries are exposed to the pressures of foreign market prices to the same extent. The USSR is less affected than its partners in the bloc; foreign trade plays a much smaller role in its economy than in the others, and in addition it is an exporter of oil and gold. The smaller countries of the Soviet bloc do not handle the threat of imported inflation from the West in the same way. Czechoslovakia and East Germany, faithful to their policy of rigid price stability, are trying to cope with the danger of imported inflation by subsidies, whereas Hungary and Poland, in accordance with their more flexible price policy, allow foreign market prices to have some effect on domestic prices. However the desire not to allow foreign market prices to have full effect on domestic prices is one of the important reasons why East European countries, particularly Poland, are indebted to the West.[75]

Up to this point we have tried to discuss the common features of the three countries' price policy. Now an attempt will be made to shed some light on the special features of this policy. The Hungarian economic reform effected in 1968 was an important milestone in Hungarian price policy. The planners recognized that to have price control it is not necessary to set all the prices from the centre. Four categories of retail prices were introduced—fixed, maximum, with upper and lower limits, and free prices[76] (the last determined by an agreement between sellers and buyers). This more flexible price formation policy was strengthened by the adoption of the rule that import prices of those consumer goods which were subject to free prices would be reflected in domestic prices. The wholesale price reform of 1968 brought the level of prices and their relations closer to retail prices, an objective which the wholesale price reform of 1959 had already pursued. Consequently the yield of the turnover tax has considerably diminished. The resulting drop in government revenues has been offset by higher direct taxes (on profits and wages) and higher social insurance contributions.

To eliminate the separation of production from consumption the turnover tax construction was changed; it has become a percentage of the price instead of a differential. In order to avoid great retail price increases, the turnover tax rates were unified only inside groups of products so that the rates are still quite numerous (1000).[77]

The reform of wholesale prices was not combined with a reform of retail prices. However, the government was determined to bring about changes in the relative prices through step-by-step price adjustments so that they would better reflect the cost of inputs and also to pass on—at least partially—increases in procurement prices to retail prices. The price adjustments of February 1966 were an introduction to this policy. Increases in procurement prices for livestock, grain, milk and sugar in January 1966 were passed on to retail prices. The price of beef was increased by almost 50 per cent, pork by 30 per cent, and dairy products by 15–19 per cent. Simultaneously prices of coal and wood for heating, and tram and bus fares increased. A compensation for these increases was given in the form of price decreases for some textile products and shoes. In addition wages were increased for a great segment of the population.[78] Similar—though not so extensive—price adjustments were repeated almost every year. In the period 1968–71 the government managed to keep price increases to an annual limit of 2 per cent, which the planners deemed acceptable. In the period 1972–6 prices grew faster, in 1976 by 5 per cent. Price increases in foreign markets, particularly the explosive price increase for oil, can bear most of the blame for the accelerated growth of retail prices.

The question can be raised: to what degree have these price adjustments brought about a change in price relativities? A. Marton indicates some very interesting figures which show—with some simplification—that price relations worsened somewhat. Prices of clothing grew faster than prices of foodstuffs. The latter can mostly be changed only as a result of explicit government decision, whereas the former are mostly determined by the 'market'. Production enterprises used their market power to obtain greater price increases than the government expected. In the period 1968–75 consumer prices increased by approximately 18 per cent; the greatest part of this increase (10 per cent) was a result of changes in free prices.[79]

In order to avoid great increases in consumer prices the government has resorted to subsidies. Till recently it managed to keep them below the level of the yields of the turnover tax. However in the last years, due to the pressures of foreign market prices, the relationship has deteriorated; while in 1974 the yield from the turnover tax exceeded subsidies by 2 per cent, in 1976 subsidies were already 3 per cent higher.[80]

In 1968 the Polish leaders decided to embark on an economic reform which was expected to go into effect in January 1970. The reform, which was supposed to keep the Polish economy within the framework of a centralised system and which also included a change in the incentive system, aimed at creating a favourable environment for the intended changes in the structure of the economy.[81] An integral part of this reform was to be a radical price reform encompassing almost two thirds of consumer goods. On the one hand, prices of foodstuffs (meat, meat products and milk), coal, building materials and also some clothing were supposed to increase, and, on the other hand, prices of durable goods and textiles were to decrease. The difference between price increases and decreases were supposed to be partially covered by increases in family allowances.[82] No wage increases were offered as a compensation. Peasants would have received compensation in the planned increases in agricultural procurement prices. The purpose of the price adjustments was to reduce subsidies on foodstuffs (most of them were not subject to price increases for many years); and, what was perhaps even more important, it was to bring about a change in the structure of consumption—to dampen the growing demand for foodstuffs, mainly meat, and to make the purchase of industrial goods more attractive. The latter was all the more desirable since the inventories of durable goods had grown.[83]

There is no doubt that the Polish price system urgently needed an overhaul. Due to the mood in the country and the way the price adjustment was prepared and announced, riots as well as human casualties and the fall of Gomulka resulted.[84] Under the pressure of these events the new leadership, after receiving Soviet loans for financing imports, withdrew the price increases (but left the price reductions untouched).[85] Earlier the new leadership had already promised to keep prices of basic foodstuffs frozen for two years.[86] Later on the freeze was extended up to 1976.[87] As for other consumer goods the government pursued a flexible price policy. In order to eliminate market disequilibria in the period 1971–6 the government carried out several price adjustments, including increases in prices of gasoline and many services.[88] In the period 1971–6 retail prices increased by 18.4 per cent.

After lengthy hesitations, in June 1976 the government deemed the time ripe for ending the price freeze on basic foodstuffs. At first it intended to increase prices of meat by 69 per cent; later it changed its mind and settled for 40 per cent.[89] At the same time it offered compensation in wage, family allowances and pension increases, and for the peasantry it offered an increase in procurement prices.[90] But again government plans were frustrated by the resistance of the population. In order to avoid greater clashes, Gierek decided to postpone the elimination of the freeze which has brought about an aggravation of disparities between procurement prices for agricultural products (which increased in 1970–5 by 36.5 per cent) and retail prices. Consequently, subsidies on consumer goods are increasing. According to M. Jagielski, deputy prime minister, subsidies in 1975 amounted to Zł100 milliard, which was 13.9 per cent of government expenditures.[91]

At the beginning of the 1960s Czechoslovakia, after abandoning the policy of price cutting, embarked on a more flexible price policy. Price stability was still regarded as a fundamental tenet of socialist economic policy; it was however interpreted less dogmatically. The idea that prices are only allowed to decline and not to go up was dropped. Stable prices do not mean unchangeable prices—this has become the new slogan justifying price increases. This new policy aimed at using prices increasingly as an instrument for influencing demand and countering evolving market disequilibria. The movement of prices was supposed to be in closer relation to economic returns, especially to the development of incomes and to the consumer goods supply.[92] Prices were also used to stimulate enterprises to extend their production of new and fashionable products. Retail price increases in the period 1961–6 were moderate; in no year did they exceed 1.5 per cent. All decisions about price adjustments were made by the central authorities.

The reform of 1966–9 with regard to prices was designed in the same way as in Hungary. In 1967 Czechoslovakia carried out a wholesale price reform with the purpose of bringing prices more in line with costs and closer to retail prices. Though wholesale prices on the average increased by 29 per cent, no changes in retail prices were made. Three groups of prices (fixed, limited and free) were introduced, with the intention of letting free prices play an increasing role along with the unfolding stabilisation of the new management system. Czechoslovakia applied the same changes in the turnover tax as Hungary. From the beginning the reformers also intended to change the price policy; they wanted to make prices not only more responsive to the market, but also to use them for stimulation of the growth of the economy. It was clear that the first goal could be achieved only if consumer price relativities were changed. They wanted to accomplish this—as the Hungarians have tried to do—by step-by-step measures.

The reformers believed that higher rates of increases are needed in order to make wages a real incentive. Faster wage increases were also regarded as important for widening wage differentials, the narrowness of which had become a hindrance to productivity growth. It was argued that problems, which a faster expansion of wages was supposed to solve, were of such importance that even moderate inflation should be accepted. At the beginning of the reform, the central authorities were reluctant to adopt this policy. They saw in it a threat to the established investment drive and were afraid that it would aggravate the existing disequilibrium in the market for consumer goods. In 1968, when the resistance to far-reaching reforms had gradually abated, the authorities reconciled themselves to the policy of faster growth of wages and its possible consequences in increased prices.[93]

After the ousting of Dubček the reformers were accused of generating inflation by allowing high wage increases and loosening the control over prices. A look at statistical figures shows that in 1968 and 1969 the cost of living index increased by 1.2 per cent and 3.6 per cent respectively,[94] which are relatively moderate rates of increase considering that Czechoslovakia went through a relatively far-reaching reform as well as a foreign occupation. The latter is partly to blame for the relatively high wage increases (8.2 per cent in 1968 and 7.4 per cent in 1969). In 1970 a virtual freeze was imposed on retail prices. The group of free prices was limited to 3 per cent of goods sold instead of the former 40 per cent. In addition, the strict regulation of prices from the centre was renewed.[95] Czechoslovakia returned to its pre-reform price policy which in essence meant that it joined the Soviet policy.

As a conclusion to this chapter it would surely be useful to give an analysis of why Poland, of the three countries which followed more flexible policies (this refers only to the 1960s in Czecholsovakia), has the worst record for inflation rates. Yet the record would be much worse were it not for the popular resistance to price adjustments in 1970 and 1976 which the government was forced to rescind. However such an analysis which could claim to be exhaustive is beyond the scope of this study. Therefore we will confine ourselves only to several comments closely connected to our topic.

Poland's inflation record is maybe the best evidence for our assertion in the Introduction that the main cause of inflationary pressure in the East has been investment overstrain. This was particularly true for the period 1949–53 but also for the period from 1958 up to now.[96] However in the latter case there is an important difference between the first part (1958–70) and the last part (1971–present).

In the period 1958–70 national income grew on the average by 5.8 per cent, but net investment in fixed assets increased by 9.2 per cent. This meant that Poland had to increase investment by 9.2 per cent in order to achieve a 5.8 per cent growth rate for national income. (Both Hungary and Czechoslovakia had a better record in this regard.) With the growing investment ratio (in 1958–70 it increased from 14.1 per cent to 21.8 per cent) employment grew by 2.8 per cent on the average,[97] much faster than in Czechoslovakia and Hungary. Poland, more than the other two countries, was under heavy pressure to find employment for its fast growing population in the working age.[98] Had it not been for the limitations imposed on employment growth, the increase would have been even higher (this refers particularly to the period 1961–5).[99]

The fast increase in employment was an important factor in the fast increase of the wage fund, faster than the planners had envisaged. In 1961–70 the share of increase in employment in the increased global personal wage fund[100] of the socialist sector was 45 per cent. The big share of employment was also due to the moderate wage increases (on the average 3.6 per cent annually). The personal wage fund increased in the period mentioned by 104 per cent, which was a 13.5 per cent over-fulfilment of the target (6.2 per cent in employment). At the same time national income increased by 80 per cent which meant only a fulfilment of the targets. However the fulfilment was not even in all the sectors. Agriculture and consumer goods industries, which are decisive for market equilibrium, did not fulfil the output targets. Agriculture only fulfilled 93.9 per cent of the target in 1961–5 and 97.4–98.5 per cent in 1966–70, and consumer goods industries achieved 95 per cent of the target in 1961–5. The fast growth of employment under conditions of moderate wage increases was surely not an incentive for high increases in productivity. So 91 per cent of the productivity plan for industry was fulfilled in 1961–5 and 101 per cent in 1966–70.[101] It is no wonder that under such conditions inflationary pressures evolved,[102] all the more because at given price relativities and level of income, the increased purchasing power was reflected in increased demand, mainly for food.[103]

In the period starting from 1971 the fast increase of investment rates had not been abandoned; on the contrary, it had been accelerated (net investment in fixed assets increased in 1971–5 on the average by 20 per cent). In 1975 net investment made up 31.7 per cent of the national income. There were however two important changes. Firstly, Poland accepted huge loans from the West to finance imports of investment goods for the modernization of its economy. Secondly, it has abandoned the policy of low wages and replaced it with a policy of fast increases. The change in wage policy acted no doubt as an important stimulus for increases in productivity and national income, and this in turn enabled higher increases in wages and investment.[104] The personal wage fund still grew faster than national income (80 per cent against 76 per cent) but the difference was much smaller than in the previous periods. Employment grew in substance at the same rate as before; however, its weight in the increment of the wage-fund declined considerably (it was 26.4 per cent).[105]

This is not to say that Poland lessened the problem of inflation; on the contrary, the rate of price increases was mildly accelerated. Though agriculture and consumer goods industries over-fulfilled planned output targets, they remained considerably behind the pace of growth of national income and wages. And this added further strains to market equilibrium and made the need for price increases for foodstuffs more urgent than in 1970.

In addition, in the period 1971-6, real wages grew at a faster rate than ever before. (They increased 46.9 per cent in 1970-6.) Due to the freeze on basic food prices, price relativities deteriorated even more during the years. With the growing level of wages basic foodstuffs became relatively more and more inexpensive. Under such conditions it is no wonder that the growing purchasing power is increasingly used for higher quality food. In the period 1970-5, consumption of meat per capita of the population increased by 33 per cent (from 53.0 to 70.3 kg), whereas in the period 1960-70 it only increased by 25 per cent.

This development has had several unfavourable repercussions. The increasing demand for foodstuffs has caused shortages, especially of meat and mainly when Poland experienced poor harvests (1974-5). Shortages push up prices on the free market and cause dissatisfaction, mainly among the lower income groups.[106] The increased demand for meat forces the government to import more and more feed grain. And this, of course, aggravates the balance of payments which is negative anyhow.

The Polish leaders face a real dilemma: for political reasons it would be advantageous to postpone major price changes for basic foodstuffs as much as possible; however, economic considerations push in the opposite direction. Even a slowdown in the growth of money wages (in 1977 to 7.0 per cent) and real wages (to 2.3 per cent) cannot eliminate the need for major price changes.

2 Wages and Inflation—A Macroeconomic Analysis

This chapter is concerned with the macroeconomic role of wage inflation in the countries under review. In view of the general definition of inflation, wage inflation is to be understood as inflation (open and repressed) caused by excessive growth of the global wage-fund in relation to the rate of increase in the volume of goods and services on which the increment in wages is to be spent. The objective of this chapter therefore is to examine the contribution of wage growth to the generation of shortages of consumer goods and services and thus to open and/or repressed inflation.

Wage inflation may result from excessive average wage increases or from excessive growth of employment or a combination of both. It is of great importance to distinguish the two factors of wage inflation since—as will be shown—growth of employment in the Soviet bloc countries was, and to a lesser extent still is, an important factor of market disequilibria.

It is clear that not every imbalance in the equation (purchasing power on the one side and value of goods available for sales on the other) can be blamed on wages, and therefore criteria are needed which would enable the role of wages to be determined. Before the criteria to be used for the analysis are identified, a warning to the reader is in order. It is not and cannot be the author's ambition (for reasons which will become clear later on) to give a clear-cut answer to the question raised above. The following two criteria are used.

1. The growth of average wages in the economy as a whole can be related to productivity growth on the same scale. It is known that

growth of average wages at the same rate as productivity is not inflationary[1] (i.e. it does not produce shortages of goods), provided all other factors remain the same. It is clear that if average non-wage incomes grow faster than average wages and, consequently, average incomes[2] grow faster than productivity, inflationary pressures will occur.

2. Actual growth of the wage fund can be related to its planned growth. This criterion results from recognition of the fact that in a planned economy the planned share of incomes including wages (minus savings) in national income is balanced with a corresponding planned supply of consumer goods and services.

WAGES AND PRODUCTIVITY

One could argue, with reference to social justice, that average wages (assuming constant prices) should grow at the same rate as productivity. As is known, under such circumstances profits, too, grow at the same rate as productivity. If there is no intention of accelerating growth and if returns on investment remain the same, profit need not grow faster than productivity for investment purposes, provided all other factors remain the same. Under such conditions it is possible to increase wages at the same rate as productivity without causing inflationary pressure.

One qualification has already been made above. Some others must now be added. First, it is assumed that gains in productivity are, at least to the extent of the increase in purchasing power, in goods which can be used directly, or with the help of foreign trade, for consumption and for which a demand exists. Of course, if the increased purchasing power cannot be matched with greater availability of consumer goods and services, inflationary pressures are unavoidable even if wages and non-wage incomes grow only at the rate of productivity. Such inflationary pressures are not unknown in a Soviet-type economy which is supply-oriented (instead of demand-oriented, characteristic of a capitalist economy) and may result from structural imbalances in productive capacities (consumer goods industries have not expanded enough) and/or from an inability or unwillingness to remedy the situation through foreign trade.[3] It is, however, questionable whether such an inflation should be classified as wage inflation. It is obvious that in the case mentioned inflation will not arise if wages grow more slowly. But it could also be argued that inflation will not arise if the supply of consumer goods grows at a pace to match wage increases. For simplification we will assume that if wages grow no faster than productivity, then it is not a case of wage inflation. This does not mean that if wages grow faster than productivity this should then be automatically classified as a case of wage inflation. (For more see pp. 43–4.)

Secondly, it is assumed that there is not only a macroeconomic matching of paid-out incomes and value of goods and services, but also a matching in the composition of demand and supply without which no market equilibrium can exist. Since in practice it is very difficult to predict changes in demand, reserves must be maintained to take care of unforeseen shifts in demand. Consequently, if non-wage incomes grow as fast as wages, then productivity has to grow faster than average wages. This is true, on the assumption that no personal savings are made. Since not all wages are spent on consumption and a part is used for savings, it can be argued that the original hypothesis (that wages can grow at the same rate as productivity without causing inflation) is correct. Finally, it is assumed that in the base year market equilibrium exists.[4]

One of the principles of wage policy that planners of the Soviet bloc countries follow is to let wages lag behind productivity; on the macroeconomic level this means behind the so-called productivity of social labour (*proizvoditelnost obshchenstvennogo truda*). The word 'social' just indicates that what planners have in mind is a macroeconomic relationship. Growth of average wages in the material sphere, including incomes from collective farms, (assuming constant prices) is related to the growth of the productivity of social labour, i.e. to national income (which in Western literature is called 'net material product') per employee in the material sphere (including agriculture).[5] This is rather an empirical rule to ensure equilibrium between growth of wages and growth of output of consumer goods and services,[6] a rule which allows large scope for manoeuvring. Nevertheless, this rule was promoted by official theory in the past as a prerequisite for economic growth.[7, 8]

Social labour productivity is not the most suitable gauge of productivity for the purpose of our study. It is limited to the material sphere, whereas an examination of wage growth impact on market equilibrium requires productivity figures for the whole economy. In addition, figures are not available on average wages in the material sphere (even without agriculture) for all the countries and periods under review.

Average wage growth can also be compared with national income per employee, which is also a productivity indicator.[9] The application of this indicator has the advantage that the denominator of the productivity formula embraces all employees including those in the non-material sphere.[10] Neither is this indicator free of deficiency. Due to the Marxist concept of national income accounting, the numerator in the productivity formula refers only to national income produced in the material sphere; the non-material sphere (services) is not considered.

In our analysis we use both sets of productivity figures—national income per employee in the material sphere, termed social labour productivity (SLP)—and national income per employee in the national economy (NIE).[11] Before these figures are examined, it should be made clear that the author does not claim that they accurately reflect economic realities in the period analysed. Too many factors stand in the way to make the measurement of national income and thus of productivity strictly objective—the analysed period being relatively long and marked by rapid economic growth, large structural changes in the economy, and sudden big changes in prices. It is known that growth rates of national income at constant prices depend to some degree on the base year used. For the comparative purpose of this study it would be most suitable if every quinquennium could have its own base year. Such figures are not available, and in addition individual countries use different base years. For all these and other reasons[12] the figures presented should be regarded as providing only a rough orientation.

In the two tables indicated, the figures display quite substantial differences. According to Table 2.1 SLP grew faster than average wages in Czechoslovakia (apart from 1961–5), the GDR and the USSR. In Hungary and Poland average wages grew faster than SLP in the quinquennium 1951–5, and also in periods of great political tension or their aftermath (1956–60 in both, 1971–5 in Poland only). According to Table 2.2 NIE grew faster than average wages in the GDR and in Czechoslovakia apart from the period of economic slowdown in both countries (1961–5). In the 1950s in the USSR average wages lagged far behind NIE. In 1961–5 there was a reversal, followed later by an almost equal pace of growth in both. Poland and Hungary (apart from 1966–70) witnessed a faster growth of average wages than NIE. The differences between the two productivity indicators are due to the differences in the dynamics of employment. In the non-material sphere employment grew much faster in the period under review than in the material sphere. This is also true of the GDR where employment grew moderately compared to other countries.

TABLE 2.1 Social labour productivity (A) and average money wages (B) (annual growth rates in per cent)

		1 1951–5	2 1956–60	3 1961–5	4 1966–70	5 1971–5
Czechoslovakia	A	5.5 (6.9)	5.8 (7.7)	1.4 (1.9)	5.5 (5.1)	(4.2)
	B	4.8	2.3	1.6	4.7	3.0
GDR	A	7.0^a (11.7)	8.5 (7.5)	4.0 3.7* (4.0)	5.5 (5.4)	(5.2)
	B	6.8	5.2	2.7	3.6	3.3
Hungary	A	4.4	4.1	4.9 4.2*	5.2	
	B	11.1	8.2	2.3	4.5	6.2
Poland	A	5.3	5.6	4.4 4.3* (4.5)	4.5 (4.1)	(7.6)
	B	12.8	9.1	3.6	3.7	9.8
USSR	A	7.7	7.2	4.8 5.1* (5.2)	6.4 (6.5)	(4.3)
	B	2.1	2.1	3.4	4.4	3.3

^{a} 1952–5

NOTES
1. All the figures in brackets refer to official figures.
2. SLP is computed from national income produced in constant prices.
3. Figures for wages are taken from Table 2.2.

SOURCES (for SLP)
Columns 1–3, *Economic Survey of Europe in 1969*, pp. 50–1.

Column 4 and the figures with an asterisk in column 3, *Economic Survey of Europe in 1971*, part II, p. 115.

Official figures: *Statistická ročenka ČSSR, 1975*, pp. 22–3; *Statistiches Jahrbuch der DDR 1976*, p. 13; *Rocznik statystyczny 1976*, p. 68 and *Narodnoe khoziaistvo SSSR 1972*, p. 63, *1975*, p. 48.

Turning to the examination of individual periods, it appears from a first glance at statistics that from 1951 and 1955 in the USSR and the GDR, average wages lagged considerably behind productivity. In that period both countries were not only free of open inflation, they even engaged in a policy of price cutting. It is, however, necessary to examine the other three countries which witnessed severe inflation up to 1953. The period 1951–5 actually falls into two periods; in the first, up to 1953 prices increased fast; in the second, 1954–5, prices declined. Now to what degree should average wage growth be blamed for this development? Tables 2.1 and 2.2 show that average wages in Czechoslovakia lagged behind productivity, whereas in the other two countries they grew much faster than both indicators of productivity. First, it should be stressed that the Polish and Hungarian figures on average wages also include a one-time, huge wage raise in connection with price adjustments; in Poland this amounted to 30 per cent in 1953 and in Hungary to 20 per cent in 1951.[13] Even if these wage increases which

TABLE 2.2 National income per employee (A)[a] and average money wages (B)[b] (annual growth rates in per cent)

	1951–5 A	1951–5 B	1956–60 A	1956–60 B	1961–6 A	1961–6 B	1966–70 A	1966–70 B	1971–5 A	1971–5 B
Czechoslovakia	5.0	4.8	4.4	2.6 (2.3)	−0.2	1.8 (1.6)	4.9	5.3 (4.7)	4.4	3.5 (3.0)
GDR	9.2	6.8	7.9	5.2	2.6	2.7	4.7	3.6	3.7	3.3
Hungary	0.9	11.1	3.7	8.2	1.6	2.3	4.9	4.5	5.0	6.2
Poland	2.4	12.8	5.2	9.1	2.5	3.6	2.4	3.7	8.3	9.8
USSR	6.6	2.3 (2.1)	4.7	2.3 (2.1)	2.0	3.7 (3.4)	4.4	4.8 (4.4)	3.1	3.6 (3.3)

[a] It refers to national income distributed in Poland and the GDR, and in the remaining countries to national income produced in constant prices.

[b] Net money wages, with the exception of Czechoslovakia and the USSR. Figures in brackets are estimates of net wages. In Czechoslovakia income tax amounted to 12.3% in 1960 and to 16% of the wage fund in 1974. We applied the following deductions; for 1956–60—11%, 1961–5—12%, 1966–70—13% and 1971–5—14%. In the USSR the tax from the population, of which the income is by far the most important, amounted to 11.6% of the wage fund in 1950, 11.1% in 1955, 9.3% in 1960, 8.5% in 1965, 9.6% in 1970 and 10.4% in 1975. (*Narodnoe khoziaistvo, 1960*, p. 845; *1965*, p. 781; *1975*, p. 743). We applied a uniform deduction of 7.5%. The estimates serve as a basis for our deliberations on Czechoslovakia and the USSR.

SOURCES

Czechoslovakia: *Statistická ročenka 1975*, p. 23, *1976*, pp. 20–3; *Incomes in Postwar Europe*, Geneva 1967, Table 7.13 (employment in 1951–5). Wages and employment refer to the socialist sector without collective farms.

GDR: *Statistisches Jahrbuch 1976*, pp. 13, 15 and 70. Employment refers to wage earners and salaried workers; wages to the state sector without the non-material sphere.

Hungary: *Magyar Statisztikai Zsebkönyv 1977*, pp. 53 and 145; *1973*, p. 38, *Magyar Statisztikai Énkönyv 1956* (employment in 1951–5, socialist sector without agriculture); *1975*, pp. 5 and 6. Employment and wages refer to the state sector.

Poland: *Rocznik Statystyczny 1976*, pp. XXXIV–XXXVII; employment and wages refer to the socialist sector.

USSR: *Narodnoe khoziaistvo v 1975*, pp. 56 and 546; employment and wages refer to wage earners and salaried workers.

were combined with even greater price increases are disregarded,[14] the fact remains that average wages grew faster than productivity.

Inflation at that time was, like that of the USSR in the 1930s, generated primarily by an ambitious industrialisation drive with all its consequences (shaped according to the Soviet concept of economic development and serving Soviet military interests to a great degree) and by excessive military expenditure. The industrialisation drive with its stress on heavy industry absorbed an increasing ratio of national income for investment[15] and brought about a dramatic change in the structure of the economy at the expense of the consumer goods industries.[16] The unsatisfactory supply of consumer goods was aggravated by the failure of agriculture to meet the plan targets, an occurrence which was primarily due to forced collectivisation.[17] Inflationary pressures were strengthened by the fact that the authorities pushed forward with the implementation of plan targets regardless of costs and efficiency.[18] This had an adverse impact on employment policy and also on wage costs.

The medium-term plans of the three countries envisaged a great influx of labour, mainly unskilled. In a sense the Polish and Hungarian planners were more realistic than the Soviet planners had been in the thirties, and relied heavily from the beginning on an expansion of employment as the main factor in implementing the output targets. They soon embraced the well-known concept of 'extensive growth'. In both countries national income growth was achieved primarily by employment growth. (In Hungary employment had a share of approximately 84 per cent in the growth of national income and in Poland 72 per cent.) Needless to say, such an employment policy laid the grounds for the hoarding of labour.[19]

The great increases in the labour force of the three countries were inflationary since they resulted in huge increases in the wage-funds which were not matched by appropriate increases in the availability of consumer goods. The situation was compounded by the fact that this great influx of labour confronted the planners with many complicated organisational problems which could not be mastered without adversely affecting productivity.

The question still remains to be answered: to what degree should the growth of average wages be blamed for inflation? It would be an over-simplification to argue that the rate of price increases corresponding to the excess growth of average wages over productivity should be attributed to increases in wages. True, in the beginning of the 1950s there was not yet in force a complete system of control over average wage and wage-bill increases in enterprises. Managers pressed for fulfilment of plan targets without due regard to labour costs, while workers pushed for higher wages due to increasing prices. Banks were lenient in their lending policy for wage purposes to enterprises. It is also true that the planners were able to control price movements more effectively than wages. Price changes could only be a result of an explicit decision by the authorities,[20] whereas wages, due to the widely spread piece-rates and progressive piece-rates[21] (mainly in Poland), could grow spontaneously. However, planners used their control over prices to bring about market equilibrium with a volume of consumer goods and services which was dictated primarily by considerations other than consumption. When, due to the worsening international situation, East European countries decided in 1950 to accelerate the industrialisation process, planners soon increased prices in order to ensure a distribution of income corresponding to investment plans. Even if wages had not gone up, planners would have increased prices in order to allow for growing investment expenditures. It should not be forgotten that this was the real reason for the decline of real wages in all the three countries.

The period 1954-5 can be characterised as a consumption period. All the countries, in an attempt to cope with the political crises which unfolded after Stalin's death, released resources for consumption purposes in order to compensate at least partially for the drop in real wages in the preceding period. In Czechoslovakia real wages increased, due mainly to price decreases, and in the other two countries mainly because of wage increases.[22]

In the period 1956-60 Poland and Hungary were again the only two countries where average wages grew much faster than productivity. This was primarily a result, as already mentioned, of the stormy political events of 1956 in both countries, mainly in Hungary. In Poland average wages increased in 1956-7 by 20.5 per cent[23] and in Hungary by even more—31.5 per cent. Interestingly enough, the huge wage increases in Hungary did not generate an open inflation. The Hungarian leadership did not dare to use price increases to depreciate the purchasing power gained. Instead it tried to cover the high wage increases by a depletion of reserves, a reduction in the investment ratio and a credit from the USSR.[24] An absolute decline in employment (in 1956 by 1 per cent, in 1957 by 1.5 per cent) helped to ease the pressure on the market. On the other hand, Poland witnessed inflationary pressures; in 1957 prices increased by 5.6 per cent, and in the whole period of 1956-60 by 2.9 per cent. As for employment, it increased in 1956 at a great rate (4.4 per cent), but later the government curtailed its tempo for the sake of fighting inflation.[25]

In 1958–60 the situation in both countries changed as they, like other countries, embarked on a new investment drive. New measures were undertaken to restrain wage increases; as a result wages grew in both countries at a slower rate than in other countries. (In 1960 in Poland real wages even declined.)

In the USSR wages continued to grow at the same slow rate as before, despite the fact that price reductions ceased. The wage reform which started in 1956 and which lasted up to 1965 brought about the tightening of work norms and an almost complete elimination of progressive piece-rates. On the other hand, it produced a rise in wage rates which resulted in a substantial increase in the average wages of some blue-collar workers, mainly in heavy industry.[26] The reform, which was gradually extended to the whole economy, also favourably influenced wage growth in the following period. The institution of a minimum wage in 1956 and its increase by stages were also beneficial for wage growth. Wage reforms in Czechoslovakia (1958–60) and the GDR (1958–60) contributed to wage growth as well.

In the period 1961–5 the relationship between wage and productivity growth appears different, depending on the chosen indicator. SLP grew faster than average wages in all the countries except Czechoslovakia, whereas NIE lagged behind wages in all the countries. (The amount of the difference between wage growth and NIE varied in different countries; it was biggest in Czechoslovakia and smallest in the GDR.) This was not due to fast increases in average wages, as they grew moderately in all the countries. It was the result of a remarkable slowdown in economic growth, though growth rates in employment (with the exception of East Germany) did not change substantially.[27] In Hungary and Poland, but mainly in the latter, there was even an increase in employment growth.

Despite efforts average wages could not be adjusted to the new situation. In a Soviet-type system growth of average wages is linked to output; with the fulfilment of increasing targets the wage-bill and average wages increase. But the growth of wages has to a certain extent its own dynamics independent of performance. There is a continuous change in skill mix towards skills requiring higher qualification, with at the same time a general improvement in qualification and also a continuous increase in the percentage share of the engineering-technical staff who are on the average better paid than the rest of the employees.[28] These facts by themselves push up average wages to a certain extent.[29] With the exception of the GDR all the countries used prices to depreciate the increases in wages. However, the price increases were moderate; in no country except Poland did they exceed the 1 per cent mark.

In the period 1965–70, in all the countries with the exception of Poland, there was an acceleration of economic growth as well as of wages. Economic reforms carried out in this period no doubt had a favourable impact on economic and wage growth. In particular, average wages grew fast in Czechoslovakia, mainly at the time the economic reform reached its peak (in 1968 8.2 per cent and in 1969 7.4 per cent). In the USSR wages reached their fastest pace in the post-war period (in 1968 they increased by 7.4 percent).[30] The ratio of average wage growth in industry to productivity growth which in 1951–5 was 23 per cent, in 1956–60 was 45 per cent, and in 1961–5 was 51 per cent, reached an unprecedented 86 per cent in the period under review.[31]

Nevertheless, SLP everywhere grew faster than average wages, whereas this was true of NIE only in the GDR, Hungary and Czechoslovakia. (In the USSR wages and NIE grew evenly.) The latter two countries managed to achieve this result despite or rather because of a far-reaching economic reform. As for Czechoslovakia, this was true only of the period as a whole; fast increases in wages in 1968–9 were inflationary. Poland and Hungary also witnessed price increases though mild ones; in the latter, they were the result of its price policy, whereas in the former wage increases were also to blame (see Chapter 1).

In the quinquennium 1971–5 only Poland and Hungary had wage increases exceeding the growth of SLP[32] and NIE. (In the USSR wages lagged a bit behind NIE.) The difference was however not big, even in Poland where wage increases were the highest. The fast growth of wages (9.8 per cent) was politically conditioned by the unrest in 1970.[33] Economically it was made possible by the fast growth of national income, though it can be assumed that growth of wages in turn acted as an important stimulus to productivity growth. As in the period 1956–8, the share of productivity in the growth of national income was very high (72 per cent).[34] This development in 1971–5 was a clear refutation of Gomulka's economic strategy which had wanted to achieve a reversal of the unfavourable situation in the economy by restraining employment and wages. Fast increases in wages due to high performance also brought about large increases in wages by explicit government decisions. When wages grow fast in many profit organisations, governments cannot avoid increases in other sectors as well. The

Polish government used four times as much funds for this purpose as in 1966–70.[35]

Like Poland, but to a lesser extent, Hungary witnessed an acceleration in wage increases in 1971–5. This resulted from a government decision to correct wage differentials by a special wage boost,[36] but it was also the result of faster price increases. Both countries, but mainly Poland, were plagued by creeping inflation which as already mentioned had its origin to a certain degree in the explosive price increases on foreign markets.

In summary, if NIE[37]—which reflects realities more accurately than SLP—is used as a basis for judgement, then it is obvious that in Hungary (with the exception of 1966–70) and Poland nominal wages grew faster than productivity. The GDR was the only country which managed to keep growth of average wages behind productivity growth through the whole period (with the exception of 1961–5). Czechoslovakia and the USSR were somewhere in the middle. Disregarding the period 1951–5 it is fair to say that when wages grew faster than productivity, the differences between the two rates of growth were, with one exception, relatively small. The one exception was the result of political tension in Poland and Hungary in 1956–7. If the question 'what caused what' (whether wages led to price increases or vice versa) is left aside, one must admit that mainly in Poland and Hungary (more in the former than in the latter) the pace of wage growth was the cause of inflation (open and repressed) in quite a few periods.

The question 'what caused what' is legitimate, mainly under conditions of a centrally planned economy. In such a system planners are able to manipulate to a great degree prices and wages according to their own considerations unless they subscribe to a rigid price stability policy. They may increase wages above productivity growth for motivation reasons and then depreciate them by price increases, the end result being that real wage growth lags behind productivity growth, which is the usual case (see Table 2.3). Or else they can increase prices to cope with market imbalances which have no direct connection with wage increases (e.g. a rise in agricultural procurement prices) and grant a compensation in the form of a raise in wages. This was frequently the case in Poland and Hungary where wages lagged mostly behind productivity. Therefore to give a definite answer to what degree wage growth caused inflation in different periods and countries, more research is needed into the reasons for price and wage changes in different periods. And this is a task which can be performed best by

TABLE 2.3 Real wages[a,b] (annual growth rates in per cent)

	1 1951–5	2 1956–60	3 1961–5	4 1966–70	5 1971–5
Czechoslovakia	0[c]	4.0	1.1	3.1	3.0
GDR		7.5	2.5	3.6	3.3
Hungary	1.0	8.0	1.8	3.5	3.4
Poland	1.9	5.3	1.5	2.1	7.2
USSR	6.8	2.8	2.5	4.5	3.3

[a] Refers to earners other than collective farmers and private peasants. Real wages for Hungary and Poland are computed from net wages and the cost of living index. Czechoslovak figures have been adjusted to net wages according to Table 2.2. Real wages for the USSR (columns 1–3) and the GDR (columns 2–3) are computed from gross wages, the remaining columns from net figures.
All figures taken from *Incomes in Postwar Europe* are compounded rates.
[b] Figures on SLP and NIE are in Tables 2.1 and 2.2 respectively.
[c] Estimate on the basis of Czechoslovak unpublished sources.

SOURCES
Czechoslovakia: *Statistická ročenka ČSSR, 1976*, pp. 20 and 21.
GDR: columns 2 and 3 *Incomes in Postwar Europe*, ch. 7, 1967, p. 34; columns 4 and 5 *Statistiches Jahrbuch der DDR 1976*.
Hungary: columns 1–2 *Incomes in Postwar Europe*, ch. 7, 1967, p. 34; columns 3–5 *Magyar Statisztikai Zsebkönyv, 1976*, p. 141.
Poland: column 1 *Incomes in Postwar Europe*, ch. 7, 1967, p. 34, columns 2–5 *Rocznik Statystyczny, 1976*, pp. xxxvi–xxxvii.
USSR: columns 1–3 *Incomes in Postwar Europe*, ch. 7, 1967, p. 34, columns 4–5 computed from figures on money wages (see Table 2.2) and the price index for physical goods (see Table 1.4).

insiders who have access to the data needed and information about the background for different decisions.

As already mentioned, inflation can be viewed as a disequilibrium due to an excess of money incomes used for consumption over the value of goods available for sale. Average wages constitute less than two-thirds of total average incomes. In order that a conclusive answer can be given to the question concerning the degree to which inflationary pressures are caused by growth in average incomes, a comparison of these incomes with growth of national income per capita should be made. The following table 2.4 gives such a comparison, but only for the period 1956–75. It reveals, with very few exceptions, the same trend as the comparison of wages with NIE.

TABLE 2.4 National income per capita (A)^a and disposable money income per capita (B)^b (compounded annual growth rates in per cent)

	1956–60 A	1956–60 B	1961–5 A	1961–5 B	1966–70 A	1966–70 B	1971–5 A	1971–5 B
Czechoslovakia	6.1	4.1	−0.4	4.0	8.8	7.5	5.4	4.6^e
GDR	9.7	7.3	3.0	2.8	5.7	3.9	5.0	4.9^e
Hungary	7.9^c	6.1^c	3.7	5.3	6.4	8.5	4.9^d	7.1^d
Poland	5.5	12.1	4.5	6.4	5.1	6.7	11.0	12.2^e
USSR	7.2	5.3	4.9	6.2	6.6	6.7	4.7	4.9

^{a} Refers to national income utilised (in the USSR produced) in constant prices.
^{b} Czechoslovak, Polish and USSR figures refer to disposable money incomes from the socialised sector (the USSR also includes incomes from the private sale of foodstuffs to the private sector), whereas the GDR and Hungarian figures refer to disposable money incomes from all sources.
^{c} Refers to 1957–60.
^{d} Refers to 1971–4.
^{e} Preliminary.

SOURCES
1. National Statistical Yearbooks.
2. S. Rudcenko, *Discussion Paper in Economics*, no. 46, 1976; D. Bronson and B. Severin, in *New Directions in Soviet Economy*, 1966, G. Schroeder and B. Severin, in *Soviet Economy in a New Perspective*, 1976, quoted according to D. Portes, *Economica*, May 1977.

WAGES AND WAGE PLAN TARGETS

All conclusions could, however, be challenged by arguing that what is of importance for market equilibrium in a planned economy is not the first criterion, which puts certain strains on the planning process, but the distributive intentions of the planners as incorporated in the plans; that is, average wages and the wage fund can be blamed for inflationary pressures even if they lag behind productivity growth and national income respectively, as long as they exceed the plan targets. It is for this reason that we have introduced the second criterion, a comparison of wage growth to planned targets in wages.

In view of the given definition of inflation, the most suitable variable for comparison is the actual growth of the wage fund for the whole economy in relation to the plan target. It is obvious that if the wage fund grows faster than the plan target and all other factors (primarily the volume of goods) grow within the limits of the plan, inflationary pressures necessarily arise. Unfortunately, figures on the plan targets for wage-fund growth (with the exception of Poland) are in such a fragmentary state as to be of no use. There are however some figures on planned targets for both components of the wage fund (average wages and employment) though, as will be shown, they are very incomplete as for average wages. A comparison of average wage and employment growth with their plan targets, instead of a comparison of the actual growth of the wage fund with its plan target, has the advantage of revealing the role of both components, which is a matter of importance for the purpose of this study.

We will begin the examination of average wage growth by first making two qualifications. Firstly, the previously mentioned 'what caused what' has its relevance even here. It is quite imaginable that an 'over-fulfilment' of the average wage plan target can be a result of a price increase which is carried out for reasons not directly connected with wages. Secondly, the comparison with plan target has its relevance if all other important factors which influence growth of average wages (and thus the wage-fund) are implemented at the rates envisaged in the plan. Otherwise the degree to which the wage plan target is met must be related to the fulfilment of the plan targets for productivity and national income. It is obvious that if the plan target for national income was over-fulfilled (without over-fulfilment of the employment target) there is good reason for an average wage increase above the plan.

The examination of the extent to which wages grew within the limits set by the plan is not an easy task, since figures are not available for the whole period. Even figures on Poland, which are the most complete thanks to the recently published book by J. Meller, do not start until 1956. For Hungary and the GDR figures are only available for the last five-year plan. In addition, these are mostly only figures for wage earners.

Figures indicated in Table 2.5 are aggregated for five-year (or shorter) periods, even in countries such as Czechoslovakia and Hungary where figures on annual plans are available. Annual figures are apparently more suitable for the purpose of the study; they are more realistic and reflect better the processes which determine inflation. On the other hand, it could be argued that the fact that annual plans were in the past often not only not finalised until the planning period was in progress, but were also sometimes changed during the planning period, reduces their comparative value. Though the situation with the five-year plans is even worse, complete annual figures for the USSR and Poland (except for 1961–75) are not available—and these are the two countries for which much more information exists than for others;

therefore figures for the longer time period are used for the sake of uniformity.

Table 2.5 confirms that average wages in Poland grew on the average faster than plan targets. The difference is considerable, mainly in the first and last quinquennia. The reasons for the first have already been explained. The difference in the 1971-5 period is probably due to the fact that in 1972, when the five-year plan was approved, the Polish planners could not predict that economic growth would take such a dramatic upturn, causing incomes to grow too. Higher wage increases

TABLE 2.5 Planned and actual average wage increases per wage earner (annual rates in per cent)

	Planned	Actual
Czechoslovakia		
1967–70	4.0	5.9
1971–5[a]	2.5–3.0	3.5
GDR		
1971–5	4.3[c]	3.7[a]
Hungary		
1971–4	5.4	4.8
Poland[a, b]		
1956–60	4.7	9.0
1961–5	3.3	3.6
1966–70	3.0	4.1
1971–5	3.6	10.3
USSR		
1959–65	3.4	2.9
1966–70	3.8	4.8
1971–5	4.1[c]	3.6

[a] Average nominal wage per employee (per blue- and white-collar worker.)
[b] Only in the case of Poland it is certain that the data refers to gross wages.
[c] Estimate on the basis of the plan for real wage growth.

SOURCES
Czechoslovakia, GDR and Hungary: Economic Surveys of Europe for the years 1967–74.

USSR: *Economic Survey of Europe in 1965*, ch. I, p. 40; *1967*, ch. II, p. 42; *1971*, part II, p. 135 and Table 2.2.

Poland: J. Meller, 1977, pp. 150, 153.

had already been envisaged in the annual plans; but they were still far from the actual increases. There was, however, good reason for average wages to grow faster than the plan envisaged due to the over-fulfilment of NIE by 21.8 per cent and national income distributed by 24.4 per cent during the five years.[38] One could argue that the growth of average wages above the target was warranted at least to the extent NIE was over-fulfilled. Otherwise, Polish figures reveal the same trend as the comparison of average wages with NIE.

The Soviet figures, however, are more in accord with the results of the comparison of average wages with SLP. They differ only in one respect, and that is with regard to 1966-70 when average wages grew faster than plan targets, an occurrence which was partly the result of the rapid growth of wages in 1968. The faster growth of wages compared to the plan target was—it seems—warranted in part by the fact that the upper range of the plan target for national income was reached.[39] When average wages lagged behind targets in 1959-65 and 1971-5, this was justified to some degree by the nonfulfilment of national income targets.[40] The few figures on other countries show that there is no great difference from what was established on the basis of Table 2.2.

As we turn to employment we embark on much firmer ground. The question 'what caused what' has little relevance here. There is no such interdependent relationship between employment and prices as between wages and prices. A faster growth of employment than planned may generate inflation, but a general increase in prices does not lead to an increase in employment, though it may lead to an increase in wages. Also, employment is not in a *direct* relationship—as wages mostly are—with productivity. On the contrary, a fast growth in employment usually reduces the rate of increase in productivity. Last, but not least, figures on planned and actual growth in employment are incomparably much more complete than figures on average wages.

In the interpretation of figures on actual growth in employment, caution is advisable. At first glance one might view such figures as an indication of the actual growth rate of the wage-fund. However, a deviation of actual growth in employment from the planned figures need not necessarily mean a corresponding deviation of the actual wage-fund from the planned. Planners, faced with a trend to an 'over-fulfilment' of employment targets, have often resorted to a curtailment of planned average wage increases. However, a greater overexpansion in employment usually generated a higher growth in the wage-fund than envisaged in the plan.

As Table 2.6 shows, in the USSR and Czechoslovakia the actual growth of employment always outstripped the plan targets. The margins were big, mainly in the 1950s (also in the USSR in 1960–5). In the last two quinquennia, the planned rates diminished and so did the margins of over-fulfilment, a result of increasing shortages of labour. In Poland the picture was mixed; in the 1950s the actual growth of employment lagged behind the targets; later on an opposite trend came into being. For Hungary and the GDR fewer figures are available. Figures for industry reveal that there was a mixed trend. As for the GDR, after the high over-fulfilment of the employment target in 1951–5 there was more under-fulfilment than over-fulfilment due to great shortages of labour. Unlike other countries, the GDR planned no growth in employment for the productive sector and an absolute decline for industry in 1971–5.

The faster growth of employment than envisaged in the plan was the result of several factors. Generally, it is possible to argue that the policy of low wages, strict control of wages, rationing of labour, the incentive system and, last but not least, over-investment played (and play) a prominent role. In different countries and periods some of the factors mentioned play a more important role than others. To begin with, is it at all possible to characterise the wage policy of the countries under review as a policy of low wages in light of what has been said up to now? Has the analysis not shown that wages grew faster than productivity in some periods? In the first subchapter we have compared nominal wages with productivity. Yet, if wages are considered in their capacity as purchasing power, which is of relevance when the question of the nature of wage policy is raised, real wages and their evolution should not be disregarded. Apart from some short periods (due mainly to political tensions or economic reforms) and in Poland since 1971, real wages grew moderately.[41] In the period 1949–53, real wages even declined in all the countries except the USSR and the GDR.

But even if we confine ourselves to nominal wages after 1955 when the industrialisation drive slowed down, we are confronted with a similar phenomenon—average wages (except for some periods) grew moderately.[42] It should be borne in mind that the absolute magnitude of average wages is small if considered in relation to national income per employee. This is also why the share of consumption of the Soviet bloc countries in national income is smaller than in other countries at the same level of economic development, though the rate of participation is higher.

TABLE 2.6 Employment (Annual growth rates in per cent)

		A		B Total state sector	
		Industry			
		Plan	Actual	Plan	Actual
Czechoslo-	1949–53	1.4	4.1	4.6 (2.7)[a]	5.9
vakia	1956–60	1.2	3.4	1.6	2.5
	1961–65	1.4	1.9	1.7	2.6
	1966–70	0.8	1.0	—	—
	1971–75	0.5	0.7	0.2[b]	0.9[d]
GDR	1951–55	3.5	5.2	2.5	4.1
	1956–60	0.6	0.2	—	—
	1961–65	0.6	−0.2	−1.2	0.7[d]
	1966–70	—	0.7	—	—
	1971–75	−0.2	0.4[b]	0.0[b]	2.6[d]
Hungary	1950–54	11.2 (8.5)[a]	9.2	—	—
	1956–60	1.7	3.2	—	3.0
	1961–65	2.5	2.3	4.2	3.0
	1966–70	1.4	2.6	—	—
	1971–75	1.5	0.2	0.9[b]	
Poland	1951–55	8.7	8.3	8.0	8.3
	1956–60	4.2	2.1	3.7	1.7
	1961–65	1.7	4.1	2.3	3.2
	1966–70	2.5	3.3	3.3[c]	3.1[c]
	1971–75	2.5	3.0	3.0[c]	3.5[c]
USSR	1951–55	2.5	4.2	2.8	4.6
	1956–60	1.9	5.1	—	5.1
	1959–65	2.9	4.5	2.9	4.6
	1966–70	2.1	2.8	—	—
	1971–75	1.1	1.4	1.0[b]	1.7[b, d]

[a] Original figures of the plan targets.
[b] Without non-material sectors.
[c] Socialist sector.
[d] Own computations which are only rough estimates due to the lack of information about what is exactly included in the planned figures.

SOURCES
Data up to 1965: *Incomes in Postwar Europe*, ch. 7, 1967, p. 11 and *Economic Survey of Europe in 1969*, part I, p. 48.

Figures for 'A' 1966–70: *Economic Survey of Europe in 1970*, part 2 (1971) p. 92; for 1971–5, *Economic Survey of Europe in 1974*, part 1, p. 96 and *Economic Survey of Europe in 1975*, p. 102.

Figures for 'B' in 1971–5: *Economic Survey of Europe in 1971*, part II, p. 115; Polish figures 1966–70 inclusive, J. Meller, 1977, pp. 122 and 150.

Small increases in wages have hampered the use of wages as an effective incentive for productivity growth. The situation has been compounded in some countries by the strict control of wages by the centre, which in practice has meant a ceiling for average wage growth in enterprises. This has tied the hands of managers in wage matters with adverse effects. For example, it has deprived enterprises of the flexibility needed in order to achieve a proper skill mix. Managers can improve the skill mix beyond the provision of the plan only if they can offset the consequent higher wages by hiring some unskilled people at lower wages. The same is true of a raise for skilled workers (who are already on the payroll) not foreseen in the plan. The Hungarian experience with hoarding labour due a ceiling on wages (1968–70) is a good example.[43]

The policy of low wages, which was reflected in wages being set after the war at a level too low to support a family, even in the most developed countries (such as the GDR and Czechoslovakia), had an impact on employment in another way. Together with ideological pressure it brought about a big influx of housewives into the labour force. Many housewives were forced to take a job only in order to contribute to the modest family income. The policy of low wages soon turned out to be a two-edged sword. It exerted pressure on women to accept employment and on enterprises to create new job opportunities.

The rationing of labour which long existed in these countries, and even now survives in most of them, is in itself an incentive to expand employment beyond the targets. People tend to hoard what is not available in the market. In addition, the direct system of wage growth regulation which mainly in the past rewarded over-fulfilment more than just fulfilment of output targets, thus encouraging over-fulfilment of plans, is also a stimulus to hoard labour. This tendency is strengthened by the desire to be able to counter frequent changes in the plan,[44] or to have sufficient labour at the end of the month and year when the workload peaks.[45] The interest of managers in having a larger labour force to manage, which is dictated not only by considerations of prestige and social status but also by material interest (basic salaries depend on the size of enterprises) pushes also in this direction.[46]

It has already been mentioned that investment overstrain is one of the main reasons for inflation (open and repressed) in the Soviet bloc. Since most of the investment projects are unique, it is difficult to predict the amount of labour, costs, and time needed for their completion. In addition, initiators of investments have a tendency to underestimate costs and labour involved in order to make the inclusion of their project in the investment plan more acceptable.[47] The situation is aggravated by the fact that too many construction projects are undertaken simultaneously, which is a source of inefficient utilisation of labour and equipment. For all these reasons it is fair to say that investment projects often require more labour than envisaged in the plan.

Finally, there is also an indirect contributing factor to the hoarding of labour—the fact that due to the relatively low costs of labour, enterprises are not interested in substitution of capital.

Of course, a faster growth of employment than envisaged in the plan differs in its impact depending on the character of the industry. If employment overexpands in consumer goods industries so that the domestic markets are supplied more plentifully, inflationary pressures may not follow. However, if the over-expansion is in branches which do not produce consumer goods, or in branches which export goods without an equivalent import of consumer goods—as has often been the case—imbalances may arise. The same is true if delays occur in putting into operation new production capacity for consumer goods.[48]

From the statistical evidence it can be concluded that except for the GDR, over-fulfilment of employment targets played an important role in over-expenditure of the wage-fund plan; this was, of course, different in different countries and at different periods. There is no way to estimate exactly the comparative share of both factors—average wage and employment—in wage-fund over-expenditure. As already mentioned, complete figures are not available for plan targets in average wage growth. What is also important is that at the present stage of research into the role of average wages, little that is definite can be said about the previously mentioned question, 'what caused what?'. What is safe to say is that, to the extent that over-expenditure of the wage-fund plan has been the cause of inflation (open or repressed), employment has played a prominent role. This statement gains in veracity when the well-known fact that employment targets were often over-fulfilled in the past is taken into consideration.

In this connection it is important to remember that, though all the countries under review have relatively effective tools for controlling average wage increases, the situation is somewhat different with regard to the control of employment growth. Not that it is more difficult to control, but rather that the authorities are more willing to condone an over-expenditure of the wage-bill in enterprises due to an 'over-fulfilment' of employment targets than the growth of average wages above the level envisaged by the plan. It is easier to make a case for an over-expenditure of the wage-bill if it can be argued that more labour had to be employed for the sake of fulfilling the plan.

Part Two

3 Regulation of Basic Wages

This chapter analyses the methods of wage control used in the Soviet bloc countries. As already mentioned in the introduction, this analysis is on a general level and comparative in nature. It is for the most part a systematisation of the methods applied or, in other words, a discussion of models of wage regulation as they exist nowadays. The analysis of the actual application of these models in individual countries is left to Part Three. (See also Appendixes 1 and 2.)

The examination focuses on the so-called *khozraschet* (profit-making) enterprises where the regulation of wages cannot be as fully ensured from the centre as in non-*khozraschet* organisations. In the latter, nominal wages and the wage-bill can grow only as a result of an explicit policy measure. Even if these organisations earn some revenues, they are obliged to surrender them to the state coffers; their expenses including labour costs are financed from the state budget. By contrast *khozraschet* organisations, for management and motivation reasons, have their finances separate from the state budget, but they are connected to it by paying taxes, by surrendering a portion of their profit and depreciation funds, and by receiving subsidies. In other words, they make decisions—the extent of which varies according to the nature of the system of management—about costs, including labour costs. Our examination is confined primarily to wage regulation in industrial enterprises.

By wage regulation we understand methods of wage-bill formation and regulation of growth of average wages and/or the wage-bill in enterprises, according to the objectives of national plans. Its purpose in a planned economy is not only to protect the economy from wage inflation but also to fulfil other important goals. As will be shown later, it has to serve as an important instrument for promoting efficiency—to put it very generally—and for influencing wage differentials in the wanted direction. Of course the two goals, protection of the economy from inflation (regulative function) and promotion of efficiency (stimulative function) may be in conflict. For more see Chapter 11.

Wage regulation can be understood in a broader and a narrower sense. It can refer to all earnings including those which are received from the bonus fund, as long as they take the form of money income; or it can be confined to basic average wages, i.e. employment incomes without bonuses from the bonus fund. The first, broader definition may be termed regulation of *employment incomes*, and the second, narrower definition *wage regulation*, referring in fact to the basic wage-bill or corresponding basic average wages. Nowadays no single system for regulating employment incomes exists; instead there are two separate systems, one for wage regulation, the other for bonus fund regulation. In a sense we can also speak of regulation of wage rates which are an integral part of basic wages.[1]

This chapter examines primarily what we understand as wage regulation, but attention is also devoted to the regulation of wage rates and to the role of Bank control and collective agreements, in brief, to all the important instruments which are used in wage control. Since wage rate regulation and collective agreements will not be further discussed in other chapters (except wage rates in Chapter 11) a brief examination of their effectiveness as tools for keeping wages within the limits of the plan is given in this chapter. The methods of Bank control and their effectiveness will also be discussed in Part Three. As for regulation of the wage-bill (or average wages) which is the main pillar of wage control, more aspects of this must first be examined in further chapters before a comprehensive evaluation of its effectiveness in shielding the economy from inflation can be given. In Chapter 4 bonus fund regulation will be discussed.

The system of employment income regulation (SEIR), as well as the systems of regulation of its components, is an important sub-system of the system of management of the economy, and it is only natural that its (SEIR) principles must be derived from the principles of the management system. Use of methods in wage regulation which conflict with the underlying principles of the system of management must necessarily undermine the system in the long run. This is not to say that a certain system of management must always be combined with the same SEIR in all aspects. As already mentioned, SEIR is an important instrument for the promotion of efficiency; therefore its concrete design (as long as an integrated regulation of employment incomes exists) as well as the design of the regulation of its components (basic wages and bonuses) is determined not just by systemic considerations. The nature of the problems faced by the economy and the intended method of their solution have also an influence on SEIR and on the system of wage regulation (SWR) which is our primary interest here.

Though the countries of the Soviet bloc have many common problems, each country has its special problems, and even common problems differ as to their extent, urgency of solution, etc. In addition countries have, for various reasons, a different scale of priorities. Thus apart from the system of management, other factors also influence decisions about wage regulation, so that it cannot be expected that countries with the same system of management will also have a completely identical SWR. This is all the more true because nowadays even systems of management in the Soviet bloc which are identical in their underlying philosophy differ in many details. Instead the system of management is rather a framework within which the SWR must act. In practice not all the countries apply a uniform SWR throughout the economy. In some sectors special conditions warrant a deviation from the prevailing system of management which may necessitate a divergence from the dominant SWR also.

From our previous considerations it is clear that the classification of the systems of management should serve as a basis for the systematisation of methods of wage regulation applied in individual countries of the Soviet bloc. In our opinion three SWRs can be distinguished in the last decade: direct (centralised), mixed, and indirect (decentralised). The direct corresponds to what is known as the administrative system of management, under which enterprises have no genuine economic autonomy and where central authorities hand down to enterprises binding output and non-output targets. Up to the reforms of the second half of the 1960s, the direct system, introduced for the first time in the USSR and later taken over by other countries of the bloc, was the dominant one. The reforms of the 1960s pushed aside the uniform system and brought into being two new systems: an indirect, a subsystem of the decentralised system of management, under which enterprises are no longer assigned obligatory targets, and a mixed, which is somewhere between the direct and indirect.

REGULATION OF WAGE RATES

Wage rate and basic salary regulation is an integral part of the control system of wages in the Soviet bloc. It would, however, be an oversimplification to assume that the only purpose of wage rate regulation is to ensure the growth of wages within the limits of the plan. Wage rate regulation is also supposed to help achieve a wanted differentiation of wages which would lead to a desired allocation of labour and stimulate workers to acquire new skills.

Until the seventies regulation of wage rates was in substance the same in all the countries. Systemic differences among the countries under review, which are reflected in employment income regulation, did not substantially affect wage rate setting. As will be shown later, only the Hungarian reform, and to a lesser degree the Polish, introduced a certain degree of flexibility in wage rate setting. Wage rates are fixed by the centre and are binding on enterprises. In most of the countries until recently, managers of enterprises had no say in determining the magnitude of wage rates. (Some flexibility existed with regard to basic salaries.) Managers, however, have always been able to influence to some degree the wage rate of individual workers by the right given to them to classify workers and jobs into certain skill grades. This right is, however, limited since the centre controls the wage-bill or average wages.

All branches of industry have at least one wage rate scale though more often several (one or more for different sub-branches). Before the wage system reform in the Soviet Union (1969–75), industry had 400 wage rate scales; now, after the reform it still has seventy.[2] The number of grades usually ranges from six to ten. In some countries and in some branches the number of grades was even larger.[3] (For technical-engineering staff and administrative workers the number of grades is much higher.)

The spread between the lowest wage rate set for the lowest grade and the highest rate (for the highest grade) is supposed to reflect, primarily, differences in the skill required for performing jobs of different complexity. There is also a difference between the rates in individual industries, which reflects the different social importance attached to them. In all the countries rates in heavy industry are higher than in light industry. The overall spread across the wage rate scales (i.e. the ratio between the highest rate in the best paid branch and the lowest rate in the worst paid branch) is much greater than within any one wage rate scale. Until recently in all the countries the wage rate for piece work was higher (7–10 per cent) than for time-work.

Job classifications are made on the basis of job evaluation manuals. Since the jobs to be performed are various and very numerous and the conditions under which they have to be performed are different, it is clear that the evaluation cannot be made from one centre. The central job evaluation manuals usually set only the rules for the allocation of jobs, as well as determining the allocation of typical jobs in individual industries to grades. They serve as a basis for drafting job evaluation manuals for individual sectors of the economy and industrial branches. Enterprises adapt those to their own needs. Naturally jobs are allocated to the skill grades indicated in the manuals on the basis of adopted criteria which vary from country to country, mainly according to the weight given to individual elements. The main criteria are the following: sophistication of the job, qualification needed for its performance, its importance or the material risk involved, and also, in some branches, the working conditions.

It would be wrong to assume that the planners are able, through the regulation of wage rates, to exert a watertight control over employment incomes. First, wage rates are a reward for work performed per hour (or shift) at normal labour intensity. Employment incomes are usually higher. Piece-rate workers receive additional payments for over-fulfilment of performance norms, and time-rate workers usually get supplementary payments. In addition workers get bonuses. If they work overtime there is an additional payment. Earnings thus depend not only on wage rates; individual effort and productivity have an impact on their level. The crux of the problem is to control the difference between total average wages and wage rates.

True, employment incomes of time-workers and salaried workers can be controlled effectively through wage rates and basic salaries. The problem lies in the control of the average wages of piece-workers who even now make up a great part of the work force.[4] (The situation was more difficult in the past in countries which used progressive piece-rates to a greater degree, such as the USSR and Poland.) The effectiveness of the control depends on the quality of performance norms. Working out so-called technically warranted norms—resulting from detailed work studies—is a time-consuming and laborious job. What is more important, only a few norms can be set from one centre for all enterprises; most of the norms are set by enterprises independently of supervisory bodies or are centrally set and adjusted by individual enterprises to their own working conditions and technological level. This circumstance gives rise to a continuous struggle between management and workers. The former strive to tighten and the latter to slacken the norms. The workers are not always the losers, though limits are indirectly set to the over-fulfilment of work norms. (In most countries the situation is regarded as ideal if the norms are over-fulfilled by about 15–25 per cent.)

The principle applied in the administrative system, that wage rates should be stable over a long period, is a further factor which reduces the effectiveness of wage rates as a tool for regulation of wages.[5] Due to productivity increases, employment incomes increase too, and since wage rates remain unchanged their share in employment incomes necessarily declines. Before the latest wage rate adjustments in the USSR, the share of wage rates in the total average wages of industry constituted only 64 per cent, in some branches even less. In the 1950s, before the reforms, wage rates declined to 40–60 per cent.[6]

Just this inflexibility in wage rates paves the way for a growing over-fulfilment of work norms. If wage rates do not change, the only way to ensure higher wages to piece-workers is to slacken work norms or to grant additional bonuses. Thus, for example, over-fulfilment of norms in the USSR in 1970–72 was 136.3 per cent in industry as a whole and 153.5 per cent in machine building.[7] The process of slackening norms weakens the purpose of norm setting. Instead of serving as an incentive for output increase and a control over wage growth, the norms become more or less a means of ensuring a certain level of income. In brief all this means that work norms, a key element in the determination of wages, is far from being under the full control of the central authorities.

Recently all the countries carried out changes in the wage rate system. These changes are marked by several common features, some of which are applied in most of the countries and some only in some countries. All the countries increased wage rates in order to make them a more effective tool for regulating wages. Consequently the share of wage rates in employment incomes has increased considerably. Most of the countries reduced the great differences in wage rates for the same work between heavy and light industry. The role of the so-called principle of the social significance of individual branches to the economy in determining wage rates was thus substantially scaled down. The tendency to smaller interbranch differentials enabled the number of wage rate scales in individual countries to be reduced.[8]

In order to encourage the acquisition of higher skills, the spread between the lowest and highest wage rates in the same wage rate scale was widened in some countries. In some branches in the USSR two supplementary grades were added to the standard six for workers who use highly sophisticated machines. In addition the wage coefficient which determines the spread between individual wage grades is higher in the upper grades.[9] In Czechoslovakia, the new uniform nine wage grade system is marked by a higher spread than the previous one.[10]

What is most important for the purpose of this study is whether changes in the system of wage rates affected its controlling functions. From available information it seems that in most of the countries no essential change occurred in this sphere. Hungary and, to a lesser degree, Poland introduced some flexibility in the wage rate system, but this has probably not added to the effectiveness of wage control. In Hungary already in 1956 enterprises were allowed to fix wage rates for individual grades within limits set by ministries.[11] However, their leeway for manoeuvring was very much limited by the strict control of average wages. (In a limited number of branches this was also applied in Poland.)[12] In 1971 Hungary further liberalised the wage rate system by introducing a uniform system for all manual workers and salaried employees. According to this system manual workers were reclassified into six skill groups, and every group into four sub-groups according to the physical arduousness of the job and to the working conditions. The wage rate of the 24 individual sub-groups was set in terms of a span ranging from 60–80 per cent.[13, 14] Recently if was decided that the wage rates applied to 24 sub-groups should increase annually in accordance with the planned increase in employment incomes.[15]

In 1972 Poland introduced a system of open wage rate scales; enterprises are assigned three to four wage rate scales with different levels of rates, and are allowed to transfer from the scale with lower rates to the one with higher rates as they achieve a higher level of employment incomes.[16]

REGULATION OF BASIC WAGES

This topic is discussed here only on a general level, devoting primary attention to systemic aspects; other important features of basic wage regulation will be examined in Chapters 5 and 6. We will start our analysis by first examining what is subject to regulation.

What is regulated?

There are three possibilities: to regulate average basic wages, or the basic wage-bill, or both. At present, regulation of wages boils down to regulation of the rate of growth of average wages or of wage-bill in comparison with the previous year. In the old Soviet administrative system both the wage-bill and average wages were directly regulated. The economic reforms of the second half of the 1960s eliminated this practice. Even in countries where the administrative system has been retained, only the wage-bill is *directly* regulated.[17]

The question can be raised: can 'what is regulated' be also regarded as a criterion for distinguishing systems? It seems that the reply can be affirmative only to a very limited degree. Both the wage-bill and average wages as objects of regulation are to a great extent neutral to the system; that is, either of them can be used by different systems. The choice of one of the two depends to a great degree on the economic and social problems to be solved. Generally speaking, wage-bill regulation by definition makes managers interested in the number of employed. Maximisation of incomes can be achieved by saving labour. On the other hand, average wage regulation by definition leaves enterprises indifferent to the problem of the size of employment. On the basis of the foregoing statements, one could argue that average wage regulation has a built-in inflationary tendency because it does not make managers concerned with the size of the wage-bill, whereas wage-bill regulation is an impediment to inflation since it induces an interest in productivity increases.

Whether wage-bill regulation will encourage labour saving and be a shield against inflationary pressures depends primarily on the design of the wage control system, that is, on the nature of the evaluation indicator, but even more on whether the indicator is a yardstick of plan target fulfilment assigned from the centre or a yardstick of performance. (Specifically, its effect on labour saving also depends on the labour reserves in individual enterprises and on how easy it is to get labour, should such a necessity arise.) The same is true of average wage regulation.

Historical experience shows that the assignment of the wage-bill from the centre and its linkage to assigned output targets makes enterprises indifferent to labour economising;[18] in many cases it even encourages expansion of employment beyond economic rationality and acts as a contributing factor to inflation. This is not to say that wage-bill regulation in an administrative system must necessarily act in the direction mentioned. Certainly it could have an opposite effect (encourage labour saving), provided enterprises are allowed to use money savings achieved from labour savings for wage increases. No doubt planners would be willing to allow such a stimulus, provided that the greatest part of the money savings would accrue to the state budget.

However, they would be very reluctant to give enterprises a great chunk of the savings for fear that such an arrangement might adversely affect product mix, might be inflationary in its consequences and might produce unemployment in some regions.[19]

Under such conditions, managers must weigh the benefits from saving labour against the benefits from over-fulfilling output targets. They can neither ignore the consequences which a revelation of reserves may have for future targets and their fulfilment, nor the fact that labour is indirectly or directly rationed. All this militates against labour saving in an administrative system.

On the other hand, average wage regulation might encourage labour saving in an indirect system. This, however, requires that the wage growth regulator be a productivity indicator computed from a net indicator such as value added, net output which reflect performance more objectively than a gross indicator such as gross value of output.

Wage-bill regulation may not only be used as an instrument for encouraging labour saving, but may also serve as a means to an indirect regulation of the size of employment. By setting the wage-bill authorities also set certain limits to the number of employable workers. In a system where the government wants to retain control over the development of employment in enterprises, the regulation of the wage-bill can be used for this purpose. This is all the more true in countries which stick to the administrative system but which, in order to make it more flexible, cease to assign employment targets to enterprises.[20]

In a decentralised system of management, enterprises must have a certain economic autonomy, and this autonomy is meaningless (or at least very limited) without the right to make decisions about employment. There is no intention of implying that a decentralised system can exist only if the central authorities do not interfere at all in employment matters. Experience shows that even a decentralised system cannot be indifferent to this aspect. Application of average wage regulation may be a stimulus to the hoarding of labour; hiring less qualified or part-time workers enables greater wage raises to be granted to workers already on the payroll.[21] The cure for these problems, however, can be found in indirect methods.[22] Some economists object to the control of employment through the wage-bill since they believe that if the control is to be meaningful, some direct control is eventually unavoidable. In comparing average wage with wage-bill regulation, J. Wilczek writes, 'In the case of the introduction of the latter (wage-bill regulation) it would be difficult to avoid the assignment of output quotas and so the reintroduction of the old system of planning by orders would threaten.'[23] It seems that this consideration played an important role in Hungary in 1971 when, despite great hoarding of labour, average wage regulation was retained as the prevailing method of wage regulation. In the Czechoslovak reform of 1966–9 the problem of whether to apply wage-bill or average wage regulation played a lesser role since—as will be shown—interference with wage growth was for some time primarily accomplished through taxes only. Taxes were levied on average wage and employment increases.

There is also another reason why a decentralised system should rather favour relying on average wage regulation. A decentralised system is in some aspects potentially more vulnerable to inflationary pressures than a centralised system (for more see Chapter 11), and average wage regulation[24]—if problems of employment regulation have been taken care of—is simpler to administer as a tool against inflationary pressures than wage-bill regulation. It also has the advantage that it is a more effective shield than wage-bill regulation against unwanted changes in wage-differentiation which may be inflationary in their consequences.

However it must be made clear that average wage regulation in itself is not a sufficient condition for classifying a system as indirect. One can easily imagine a combination of an administrative system with average wage regulation. This was the case in Hungary during 1957–68 when the government assigned enterprises binding maximum rates of wage growth.[25] Hungarian planners resorted to this method of regulation just for the purpose of countering the wage-push generated by the stormy events of 1956. Similarly the first Czechoslovak attempt to reform the system of management of the economy in 1958 also meant an introduction of average wage regulation in accordance with growth of productivity.[26] In both cases, however, average wage regulation was introduced in a situation where an effort was being undertaken to make the system of management a bit more flexible. What really matters for classifying systems in the final analysis is not what is regulated, though this plays a certain role as has been shown, but how wages are regulated: whether directly or indirectly.

It will be interesting to follow the effects of the Hungarian changeover to wage-bill regulation. In the last two years they increased the weight of wage-bill regulation and made it to the leading operational rule in 1978. According to J. Lökkös average wage regulation can work relatively well under conditions of an extensive type of growth whereas in a transitional stage (from the extensive to the intensive type) it will only work with constant corrections. 'In the intensive stage of development the negative features of the average wage regulation prevail'—writes Lökkös—'more and more impeding the efficiency and rational economy of labour.'[27] In other words, he implies that if economic growth is based primarily on an increase in productivity, wage-bill regulation should be used. This statement has the shortcoming that it considers 'what is regulated' in isolation from many important factors. The experience of Hungary up to now does not support his statement unequivocally. In addition the Hungarian system is not a pure wage-bill regulation. Once the average wage increases to a certain level owing to an expansion of the wage-bill, its further growth is regulated by taxes. The Hungarian planners are trying to achieve two goals with this combination. Wage-bill regulation is a new attempt to induce enterprises to adopt a more rational employment policy, whereas the built-in average wage regulation should take care of possible wage inflation.

How is the wage-bill formed?[28]

To begin with, it should be mentioned that in the direct and mixed systems the formation of the wage-bill does not have a separate existence from its regulation. The two processes coalesce. Assignment of the wage-bill from the centre is at the same time a regulation of the wage-bill to a great degree. In the indirect system, on the other hand, the formation and regulation of wages are clearly separated. The first is more or less under the jurisdiction of enterprises and the second of the central authorities. Experience up to now has revealed the application of two methods: (a) the wage-bill is planned (assigned) from the centre or (b) it is left to enterprises to determine the wage-bill.

The assignment of the wage-bill from the centre may take one of two forms: it may be assigned as an absolute or a relative sum. The latter means that the size of the wage-bill as a whole or its increment over the previous year is given in a certain proportion to the fulfilment of an assigned success indicator (or indicators). Such a system includes a normative which relates the size of the wage-bill (its increase over the previous year) to the degree of fulfilment of the success indicator.

The first solution (the assignment of the wage-bill as an absolute sum) can be classified as a direct method—the Soviet model is the classic example. In the Soviet model, which was more or less used by the other countries up to the second half of the 1960s, the centre assigns enterprises the wage-bill as an absolute sum. This can be termed—the planned wage-bill. In fixing this sum the centre takes into account the output target, the planned number of employees and their qualification mix, and allows for average wage increases depending on the planned growth of productivity. The actual size of the wage-bill is usually different from the planned size. It can be the same if enterprises exactly fulfil the targets of the plan. In the case of over-fulfilment or under-fulfilment of plan targets expressed in an output indicator, usually gross value of output,[29] the actual wage-bill increases or decreases in relation to the planned amount. The change in the wage-bill is now no longer proportional to the degree of fulfilment of the target; the adjustment coefficient is smaller than unity.

The East German system in its present form does not differ substantially from the Soviet system. The only important variation is that in East Germany the wage-bill assigned to enterprises can be used as long as enterprises fulfil the assigned target in productivity. Not every over-fulfilment of the plan targets entitles enterprises to a higher wage-bill. The size of the increase is determined by management bodies above the level of enterprises.

The second way of assigning the wage-bill cannot be classified without taking into account the way it is linked to evaluation indicators.[30] If the planned wage-bill is assigned as a fixed proportion of a planned (assigned) success indicator, we regard such a system as a direct one. The present Czechoslovak system may serve as an example. In it the planned wage-bill is assigned as a percentage of the plan target for marketed output. The actual wage-bill increases or decreases compared to the planned if the plan target is over-fulfilled or under-fulfilled. The differences between the Soviet and the GDR systems on the one hand, and the Czechoslovak on the other, are more of a technical nature.

If the growth of the wage-bill depends on the degree of fulfilment of the success indicator over the previous year, we regard such a system as mixed. The new Polish system (which was launched in 1973, and which, according to plans, was supposed to become the predominant system by 1977) fits this characterisation. In this system the growth in the wage-bill over the previous year is linked to the increment in output added over the previous year. For every percentage point of increase in output added, the wage-bill increases by a normative assigned from the centre.

To sum up: the difference between the direct and mixed systems is that in the former case the centre fixes the amount of the wage-bill (the planned wage-bill) which enterprises will be allowed to use if they fulfil the assigned targets. For an over-fulfilment, additional funds are granted. In the latter case, the centre does not really assign the wage-bill; what it assigns is the success indicator which is to be used as a yardstick for measuring performance, and the normative, which relates increases in the wage-bill to performance. The size of the wage-bill (instead of being linked to the fulfilment of output targets) thus depends on the degree of fulfilment of the evaluation indicator over the previous year and on the normative.

In the indirect system it is left to enterprises to determine within the framework of central regulations how much of the resources resulting from their market performance will be used for wages; this was the case in the Czechoslovak reform (1966–9) and with some qualifications in Hungary up to 1971. This gives enterprises a fair amount of authority in the distribution of the gross income obtained from selling their products. On the other hand, enterprises take over the responsibility for generating the funds needed for remuneration.

How are the wage-bill and average wages regulated?

With this we have arrived at the most important criterion for distinguishing systems. Planned economies of the Soviet bloc cannot do without central regulation of wage growth. They can, however, carry out this regulation directly or indirectly.

The direct method of wage regulation has already been more or less described in the preceding pages. It consists of linking the wage-bill to a centrally planned success indicator. The wage-bill is assigned by the centre; its actual size is allowed to be higher or lower than the planned size, depending on the degree of fulfilment of the assigned success indicator. In this way the central authorities try, on the one hand, to use wage regulation as a stimulus for over-fulfilment of targets and, on the other hand, to keep wage increases within the limits of performance.

The opposite of this method of regulation is the indirect method implemented by way of taxation. Increases in average wages over the previous year are subject to taxes—progressive or non-progressive—which put limits to possible wage growth. It is left to enterprises to make decisions about growth of wages; however they must weigh their decision in the light of taxes to be paid.

The Czechoslovak system under the reform of 1966–9 can be regarded as relatively the purest representative of the indirect system. The Czechoslovak reform abolished assignment of the wage-bill from the centre and made its growth a function of performance which was measured by gross income. Since there was no separate bonus fund regulation, the wage regulation encompassed wages and bonuses at the same time and was achieved through taxation for some time.

The Hungarian SWR went through several changes. In the period 1968–70 wage increases were financed from the bonus fund which was fed from produced profit and regulated by taxes. In the quinquennium 1971–5 average wage increases were linked to gross income per employee in the vast majority of enterprises, while nowadays wage-bill regulation (as the most important method) is linked to value added. Though taxation plays an important role as a regulator of wage growth, it was and is only a supplement to the linkage of wage growth to an indicator of performance. This double safeguard proved, as will be shown later, to be a good shield against possible wage inflation, definitely superior to the reliance on taxation only. It is justifiable to raise the question whether the Hungarian SWR, in its existing form, should be classified after all as an indirect system. There are several factors which put the Hungarian system close to what we call an indirect system; however, for methodological reasons, it would be better to discuss first the mixed system.

Between the two extremes of wage regulation mentioned, a compromise solution exists which may be denoted as mixed. On the one hand, the growth of the wage-bill is dependent on performance, i.e. on the degree of fulfilment of an evaluation indicator related to the previous year—and thus not on the fulfilment of a planned success indicator, usually a gross output indicator, as in the direct system.

This change is of importance. The linkage of wage-bill growth to centrally planned output targets stimulates enterprises to fulfil and over-fulfil output targets, regardless of the situation in the market. This fact alone can be a source of market disequilibrium. There are also other factors which strengthen this trend. In order to maximize the wage-bill, enterprises look for easy ways to fulfil and over-fulfil targets. As will be shown later, gross value of output as an indicator offers a good opportunity for this. Distortions in price systems are also helpful in this regard to some branches of industry and to some enterprises. Finally the additional allocations to the wage-bill for over-fulfilment of output targets may be a destabilising factor. Furthermore, this system encourages enterprises to conceal their real productive potential and reserves. Acceptance of more stringent plans becomes an impediment to easy fulfilment of assigned targets.

The transition to a linkage of the wage-bill to performance over the previous year aims at overcoming some of these drawbacks. It cannot, however, prevent enterprises from looking for 'easy' ways or from taking advantage of price distortions in order to accomplish higher rates of performance. Yet non-direct systems are in a better position than direct ones in this regard since they do not use gross value of output as an indicator. (See further, Chapters 5 and 11.)

On the other hand, the centre fixes the rate (normative) at which the wage-bill can grow with the increase in the success indicator, somewhat like the direct system; thus, regulation is not done by taxation as in the indirect system.

For those who might maintain that the difference between the mixed and the indirect systems is negligible, and who argue that both systems depend on performance, the following points should be mentioned. First, in the case of regulation by a normative, the centre makes the decision as to how the gross income will be distributed; whereas in the indirect system, where wages are regulated by taxation, enterprises themselves make the decision as to how gross income will be distributed, naturally taking into consideration tax regulations. The centre's interference with enterprises is also greater in the first case because the normative is usually not uniform, whereas tax rates usually are.

This brings us to a second difference which lies in the potential impact of both regulation systems on economic efficiency and labour economy. In the indirect system enterprises must earn funds for wages, and taxation can be used as a criterion of and stimulus to efficiency. (For more see pp. 94–5). In the mixed system, if an enterprise fulfils the size of output of the previous year (not to say if it exceeds it) it may have enough funds for wages though it may be at a level of efficiency below the average. The indirect system, if implemented consistently, may be a stimulus to a more rational utilisation of labour whereas the mixed, viewed simply in systemic terms, lacks such a stimulus.

Finally, an indirect SWR can only be a product of a decentralised system of management which, in turn, can only come into being if elements of the market mechanisms are introduced. If the centre ceases to assign targets to enterprises and to coordinate their activities *directly* the market must take over this role. Under such conditions, the wage-bill is the result of performance, determined to a great degree by market forces. What is also important is that the stimulative effect of the desire to maximise the growth of average wages is provided by the market mechanism.

The mixed SWR does not presuppose the existence of elements of the market mechanism.[31] In this system the normative performs two functions: the stimulative function which the market fulfils under the indirect system and, in addition, the regulative function which is fulfilled by taxation in the indirect system.

What has been said about the mixed system fits, as already mentioned, the Polish system. A short comparison of the Polish and Hungarian systems will show why the Hungarian one is closer to what we termed the indirect system. In both systems growth of wages is linked to performance over the previous year (in Poland measured by output added, in Hungary by gross income per employee, or value added). This distinguishes the systems in both countries from the direct system.

In the Hungarian system taxes are also used as a regulator of wage growth. Wage increases above 6 per cent or increases above the limit to which enterprises are entitled on the basis of performance are subject to taxes. The Polish system of 1973 did not provide for such a possibility. Only recently has Poland introduced a system of contributions (charges) to the branch (ministry) reserve fund which is intended to be an additional instrument of wage control. If the actual implementation of the two instruments is judged from a systematic angle, it is safe to say that there is quite a difference between them. In Hungary taxes for wage increases are, as a matter of principle, applied equitably; the trigger threshold for payment of taxes is the same for all enterprises and the tax rates are the same for the same form of wage regulation. All these rules of the game are spelled out in advance and are supposed to be in effect for a long period of time. In Poland the trigger threshold for contributions as well as the contribution rates are not equal for the associations converted to the modified system in 1977, and in addition, they are subject to annual changes. (See further, Chapter 8.)

In summing up it is possible to contend that while taxes in the Hungarian system are an indirect instrument, contributions in the Polish system due to their conceptualisation are far from what is understood as an indirect method. Up to 1976 the Hungarian system also differed from the Polish in that the normative for average wage increases with improvement in performance was, with some exceptions, long term and uniform.

BANK'S CONTROL FUNCTION

Most of the countries do not only rely on the system of wage and wage rate regulation which has been discussed. They also apply administrative controls which aim at discouraging enterprises from wage overexpenditures. Apart from administrative control by supervising organs, enterprises are subject to the control of the State Bank. The control of the Soviet Bank over wage expenditure developed gradually out of experiments in the 1930s.[32] Later changes, even during 1965, have been mostly of a minor nature. In the beginning of the 1950s Bank control was taken over by the other countries under review. The reforms of the 1960s brought about changes in this sphere; the control function of the Bank over wage expenditure was, in substance, replaced in Czechoslovakia and in Hungary by indirect control by taxes.

Bank control over wage expenditure is part of a comprehensive package of controls over the economic performance of enterprises known in the literature as 'control by rouble' (mark, crown, etc.). This involves not only financial supervision but also a regular check on enterprises to verify whether their financial operations conform to their plans. For this purpose all enterprises are obliged to make all payments for goods and services through the local branch of the State Bank as well as receive payments through the same channel. Commercial credit is prohibited; enterprises can receive credit only from the local bank branch. With some simplification it is possible to say that enterprises use cash only for payment of wages; all other operations are book transfers on their accounts with the bank.[33] The carrying out of all financial operations through the monobank is supposed to give the latter a good insight into the enterprise's position with regard to fulfilment of plan targets. Yet the monobank's function is supposed to be an active one; the bank's access to information should facilitate the performance of one of its main functions, to make enterprises keep their expenditures within the limits of the plan. In this regard the control over wage expenditure is—due to its importance for economic growth and market equilibrium—of greatest significance.

The monobank is most suitable for this function since, in contrast with the supervising bodies of enterprises, it is not directly interested financially in the plan fulfilment of enterprises; the bonuses of the managers of the monobank do not depend upon the performance of enterprises.[34] The monobank has at its disposal a certain number of instruments which have to ensure that enterprises' disbursement of wages is within the authorised limits. Without going into details, it can be said that, on the one hand, it can confine itself to a report to supervising bodies about the enterprise's wage over-expenditure and rely on its taking some corrective measures, or, on the other hand, it can itself apply sanctions. One of the ways in which supervisory agencies may penalise managers for over-expenditure of the wage-bill is by suspending a part of their bonuses. This is what the rules allow in East

Germany,[35] in the USSR[36] and even in Poland.[37] It seems that in Czechoslovakia only the bonus fund is affected and not the managers' bonuses as such directly.[38]

If the overdrafts are consecutive and large, banks may in some countries refuse to give funds above the authorised limits. To what degree banks use their power it is difficult to say. To withhold funds for wage payment for the workers of an enterprise is not just a legal decision; it may have far-reaching political repercussions. Therefore it can be assumed that the monobank, without the blessing of supervising and political organs, would not dare to undertake such a step.

COLLECTIVE AGREEMENTS

Collective agreements between the management and the Trade Union organisation of enterprises are also one of the government's tools for regulation of wages.[39] For a better understanding of the role of collective agreements, a few words must first be said about the role of Trade Unions (TUs) in the Soviet bloc countries.

There is no agreement among economists about the role the unions have played in generating the severe inflation which plagues Western economies. Some economists see in the behaviour of unions the main cause of inflation.[40] Others attribute a much smaller role to them, and there are, of course, authors whose views vindicate the trade unions completely. Needless to say, all these views are much influenced by *Weltanschauung*. One thing that is abundantly clear is that the determination of wages in enterprises where workers are unionised is a result of bargaining between management and unions. Trade unions are a force which sees to it that employees receive a reasonable share in productivity gains. The fact that real wages in the West do not lag as much behind productivity growth as in the East is, no doubt, an accomplishment the TUs should be credited with.

In the Soviet bloc countries protection of the material interests of the population is a secondary task for the TUs. Their primary task is, to put it briefly, to rally support for the Party policy. In the economic sphere where their activity is the most important, this means helping to mobilise the masses for the fulfilment of economic goals set by the national plans. And this includes also supporting and rallying support for the centre's wage policy. This is not to say that TUs have no input at all in the process of decision-making about wage policy. They are at least consulted when decisions are made. However the TUs are not an autonomous force which can act and use their power in any way they see fit in accordance with the interest of the workers they represent. The leaders of TUs owe their positions to the Communist Party, and if they want to retain them they must act according to its instructions.

This is not to say that central authorities can disregard the material interests of the workers. It is possible to agree with R. Portes that the population's expectations of an annual wage increase (generated to a great degree by Party propaganda) exert pressure on the central authorities.[41] But these pressures are, no doubt, smaller than in the West. In normal periods the authorities can manipulate people's expectations a great deal. The centrally directed media are of great help in this respect. Of course the situation is different in periods of political tension. As has already been shown, the greatest wage increases were usually achieved in periods of political tension or just afterwards.

From what has been said it is clear that collective agreements in the examined area cannot be a result of free collective bargaining by two independent parties in which each, using its bargaining power, is trying to give the least and to receive the most. In this regard there is only a minor difference between the individual countries of the Soviet bloc. The contents of collective agreements are primarily an adaptation of regulations and rules of government wage policy to the actual conditions in individual enterprises. 'Regulations and rules (of the collective agreements),—writes a senior official of the Hungarian TUs—'should not be in contradiction with higher level legal provisions'[42] (rough translation). 'Collective agreements should also be used for the regulation of wage growth in individual organisations' is one of the provisions contained in the Czechoslovak Ministry of Labour's announcement about the regulation of wages.[43] If legal provisions or regulations change, collective agreements change, too, to reflect the change. Or the new central regulations are implemented even though they are not in accordance with the collective agreements concluded.[44]

Here it is worthwhile mentioning that even the Czechoslovak reform, up to the second half of 1968, viewed collective agreements in the same way. The government decree on the General Conditions of the Economy and Management of Enterprises, which contains the principles of the new economic reform, classified collective agreements—like the wage-rate system—as an instrument of direct regulation of wages.[45] Collective agreements concluded at that time contained the government provision that investment funds and other funds should grow more rapidly than the wage-bill. When in 1967 the government supplemented the regulation of wages by taxation by a linkage of wage increases to productivity gains, the collective agreements were simply rewritten in order to meet the new requirements.

Collective agreements deal not only with remuneration, distribution of bonuses, working conditions, in brief with the material interest of workers.[46] They are also an instrument for the fulfilment of the plan targets. According to one Hungarian author, collective agreements serve one aim—'increases in output and economic efficiency. . . .'[47] They contain details about the workers' contributions to the fulfilment of the plan and the corresponding remuneration. For this purpose enterprises' plans and collective agreements are prepared and discussed by their employees at the same time. This also serves the purpose of giving workers the feeling of being part of the decision-making process.[48]

In what sense do collective agreements help to cope with inflationary pressures? Needless to stress, they are surely not the most important pillar of the wage control system. Their importance is in their aim, which is to gain workers' active or at least passive support for the applied policy of wage regulation and in this way to induce them to a moral commitment to wage restraint. The structure of the collective agreement is tailored to this end. Projected wage increases and the distribution of bonuses are conditioned by the fulfilment of certain economic targets. Such a linkage serves, on the one hand, as a stimulus to greater effort and, on the other hand, as an argument against excessive wage increases. The significance of collective agreements is surely enhanced by the fact that the management must commit itself to improve working conditions, catering, care of children and of working mothers, housing, etc.

It is difficult to assess to what degree the collective agreements fulfil the above-mentioned functions. One thing that seems to be clear is that many workers do not perceive the collective agreements as a reflection of their interests. For example a public opinion poll carried out in Czechoslovakia showed that 34.2 per cent of respondents dared to contend that collective agreements only partly, and 3.1 per cent only very little, reflect the interest and needs of the workers.[49]

4 Regulation of Bonuses

In the preceding chapter, regulation of basic wages was discussed. However, employees of enterprises also receive bonuses, premiums and rewards from funds outside the wage-bill,[1] mainly from the incentive fund. Even if these bonuses[2] are not very large—their present range in the Soviet bloc countries is from 5-12 per cent of the enterprise wage-bill including bonus fund[3]—they must be regulated in order to ensure that their magnitude does not exceed the plan limits and thus contribute to inflationary pressures. They must also be regulated since they serve as the main incentive,[4] a function which can be filled only partially by basic salaries (and the same is true of wage rates), which are fixed by the centre and in most countries changed by it at long time intervals. The size of basic salaries is primarily supposed to be a reward for knowledge and experience gained over a long period. It is felt that it would be wrong to change them in accordance with short-term fluctuations in performance. In addition it would be politically difficult to reduce basic salaries (or wage rates or even basic wages) in the case of nonfulfilment of targets.[5]

In this chapter we will discuss the basic common and contrasting features of the incentive systems, particularly with regard to the formation and regulation of the bonus fund, as they have developed in the last decade in the countries under review. Due to the fact that bonuses are destined more to top managers[6] who are the moving force of economic activities, special attention will be devoted to their incentives. The chapter ends with a short survey of the role of bonuses in employment incomes.

In all the countries (with the exception of Czechoslovakia in 1966–9 and Poland[7]) the reforms have brought into being a single incentive fund,[8] which is separate from the wage-bill and which is supposed to provide the bulk of the bonuses. The way the bonus fund is administered gives some indication of the nature of the system.[9] An integrated wage-bill, in the sense that the centre confines its interference to a regulation of employment incomes as a whole, is characteristic of an indirect, decentralised system. Only such a system can afford to allow enterprises to determine (naturally within the framework of general regulations and constraints) how much of their resources will be used for employment incomes, and of these how much for basic wages and salaries, and how much for bonuses.

An integrated wage-bill is contrary to the spirit of an administrative system. It means a delegation of decision-making power in a very important area which an administrative system can hardly afford. To want to have everything under strict control is a basic feature of such a system. Apart from this, there are also other reasons why the centre is interested in having the right to fix the wage-bill and the bonus fund separately. The planners are afraid that enterprises may abuse their right to distribute the integrated wage-bill and commit themselves to higher basic wages and salaries for some groups than the centre would approve of. This might lead to a decline in the weight of bonuses which would weaken incentives and might also contribute to the generation of inflationary pressures. As is known, workers prefer increases in basic earnings to equivalent increases in bonuses.[10] Increases in the fixed components of wages mean a commitment by enterprises for future years, whereas increases in the variable part, such as bonuses, do not.

We have mentioned that an integrated wage-bill is only compatible with an indirect system. However this is not to say that an indirect system or a system close to what we call indirect must necessarily be combined with an integrated wage-bill, as Hungarian experience proves. On the other hand, an indirect system without an integrated wage-bill weakens the decision-making of enterprises in an important sphere.

The fact that the bonus fund is regulated differently from the wage-bill can be characterised as a two-channel system. Viewed from a systemic angle, the difference between the regulation of the bonus fund and the wage-bill cannot be very great since both are subordinated to the existing system of management of the economy. Yet some systemic differences do exist. For example, the Soviet incentive system (1966–71), like the German during 1969–71, was more in line with what we call a mixed system, whereas the regulation of wages remained direct in nature. The Hungarian regulation of the incentive fund is indirect whereas basic wage regulation is less indirect. This is mainly true of the new SWR introduced in January 1976.

In all the countries profit is the success indicator (or one of the indicators) which regulates or helps regulate the size of the bonus fund. As will be shown later, profit has not played the same role in all the countries; its application is influenced by systemic differences. On the other hand, growth of the wage-bill, as already mentioned, is linked to a gross output indicator in a direct system, whereas in non-direct systems it is linked to a net indicator which is at the same time a sales indicator.

What is the rationale for planners in a centralised system to opt for such a combination of indicators in the two-channel system? The answer seems to be clear. Such a system rests on the belief that only strict control over production through binding targets can induce enterprises to perform well. Apart from providing a yardstick for wage increases, the linkage of wages to a gross output indicator has to induce enterprises to maximise output. The use of profit or some other synthetic indicator as an indicator for the regulation of bonuses has to protect the economy against the tendency built into quantitative indicators, namely, that enterprises are encouraged to expand output without being forced to take proper account of costs, quality and demand.

A two-channel system has the advantage over an integrated system that it allows more aspects of economic activities to be linked to incentives, thus intensifying the control, a circumstance which is appreciated in a centralised system. On the other hand, a two-channel system can work reasonably provided that the two sets of success indicators are complementary and not conflicting. Needless to stress, it is not an easy task to design indicators in such a way. (See further, Chapter 5).

It can only be speculated why Hungary resorted to the two-channel system. In the first stage of the reform it seemed that it would follow the Czechoslovak pattern in this regard though in modified form. Their change of policy may have resulted from the recognition that regulating wages by profit as was done in the first stage was unsuitable.[11]

FORMATION AND REGULATION OF THE BONUS FUND

The reforms of the second half of the 1960s brought about changes in all the countries in the formation and regulation of the bonus fund, which is the most important component of the incentive system. In the period which has elapsed since the reforms, the incentive systems have undergone further changes, of different magnitude in different countries. If the same systematisation scheme is used as for basic wage-bill regulation, three systems of incentive fund regulation can be distinguished: direct, mixed and indirect.

The bonus fund can be fixed from the centre or determined by enterprises; at least this has been the practice thus far. In the first case the bonus fund may be assigned as an absolute or relative sum. If it is assigned as an absolute sum, it can be regarded as a direct system, which is characteristic of the present systems used in both the Soviet Union and the GDR. In the Soviet Union the bonus fund (the so-called fund for material stimulation) is assigned to enterprises as an absolute sum directly by the supervisory ministry or, more commonly, through the association. The bonus fund of an enterprise is a portion of the bonus fund of the association (within whose jurisdiction the enterprise in question lies), and this in turn is a portion of the bonus fund of the ministry. With some simplification it is possible to assert that for the two quinquennia (1971–5, 1976–80), the departure point for setting the funds of ministries, associations and enterprises was their planned wage-bill for the last year of the five-year plan. In setting the size of the funds two criteria were used: the economic importance of the branch and enterprises respectively and the weight of technical engineering and administrative staff in the number of employees.

In 1971–5 the planned annual growth of the bonus fund was linked to gross value of output or commodity production.[12] For each planned increment of one rouble in gross output, the incentive fund grew by a constant fraction of a rouble through the years (long-term normatives).[13] In the present five-year plan the annual growth is linked, according to A. Miliukov, to the increment in sales-revenues achieved due to an increase in productivity.[14]

Ministries and associations are allowed to use up to 10 per cent of their funds earmarked for bonuses for a centralised reserve fund. Its purpose is to contribute to a stabilisation of normatives; it can be used for allocations to the bonus funds of enterprises and associations which have suffered a temporary decline in the fulfilment of targets due to the introduction of new technology, or of enterprises which produce consumer goods of low profitability.[15]

So far, only the first stage of planning of the bonus fund has been considered. In the second stage, enterprises are directly involved. If they voluntarily accept higher targets, the planned bonus fund increases on the basis of stable normatives fixed differently for each success indicator. At present the ministries are allowed to assign to associations and enterprises three or a maximum of four indicators; increase in productivity and increase of the percentage share of high quality goods in the total output should in principle be applied everywhere,[16] as the new incentive system puts great stress on productivity (to offset increasing shortages of labour) and improvement in quality. Other indicators which could be applied are the following: gross output (in physical or value units) or profit or reduction of costs of production, etc. In other words, profit is no longer a binding indicator.

The actual size of the bonus fund depends on the fulfilment of the success indicators; if they are fulfilled, the real size of the incentive fund corresponds to the planned. In order to encourage enterprises to accept realistic plans (but at the same time demanding ones), over-fulfilment of the plan is rewarded by at least a 30 per cent lower normative. Under-fulfilment is penalised in the opposite way. The incentive fund also has maximum and minimum limits. In the quinquennium 1971–5 its size as a percentage of profit was not allowed to exceed the level of 1970. The minimum limit is 40 per cent of the previous year, provided, however, that the enterprise has sufficient profit to finance the incentive fund.[17]

The present GDR system is similar to the Soviet. Starting with 1971 the centre, through the association, has assigned the bonus fund to enterprises as an absolute sum. In 1971 the actual size of the bonus fund depended on the fulfilment and over-fulfilment of the planned net profit and, in addition to this, of two supplementary material targets.[18] For 1972 a modified system was introduced which, with some minor changes, remained in force also in 1973 and 1974. The planned bonus fund—assigned as before as an absolute sum—could increase if enterprises voluntarily accepted higher targets in profit and commodity production at the stage of plan drafting.[19] However, the allocations to the bonus fund for the over-fulfilment of the second indicator could be made only if two additional targets were fulfilled. Among the targets, export deliveries and the supply of the domestic market figured prominently, as in 1971.[20] In 1974 over-fulfilment of voluntarily accepted targets was rewarded by a smaller normative than fulfilment of planned targets.[21] The same incentive system, with some small modifications as to the normatives, is in force for the present five-year plan period (1976–80).[22]

Reference has already been made to the fact that the bonus fund fixed by the centre can take the form of a relative sum. This is understood to mean that the bonus fund as a whole, or its increment over the previous year, is given in a certain proportion to the fulfilment of the success indicator. In that case the nature of the system is determined by the way the bonus fund is linked to the success indicator. If it is assigned as a *fixed* proportion of a target *planned* by the centre, the system can be regarded as direct.

The present Czechoslovak system is such a system. Since 1971 the bonus fund has been assigned to enterprises by their supervisory bodies as a certain proportion of planned profit.[23] This proportion (normative) varies for different enterprises. Under certain conditions the bonus fund can be assigned as an absolute sum.[24] Up to 1974 the bonus fund was calculated from 'adjusted profit', that is, the amount which resulted from dividing profit by the index of rate of growth of the wage-bill compared to the previous year. Apparently this method of calculating the bonus fund was motivated by the fear that profit as a success indicator might lead to the expansion of employment, since the simplest way to increase the amount of produced profit is to expand output by hiring more labour. At the same time it had to encourage restraint in wage-bill increases. However, it turned out that this operating rule had only a small effect—due to the smallness of the bonus fund—on the behaviour of enterprises; in addition it also penalised enterprises whose higher wage increases might have been the result of more efficient performance, and for these reasons it was dropped.[25]

The bonus fund of enterprises amounts to its planned size provided the planned target for profit is fulfilled. Enterprises could achieve further allocations to their bonus fund if they voluntarily accepted a higher target for planned profit. For the fulfilment of the voluntarily accepted target the allocations were higher; starting with 1973 they were twice as high as for the original target. A failure to fulfil the increased target was usually penalised more heavily than a failure to fulfil the original plan.[26]

With the start of the sixth five-year plan (for 1976–80) some provisions for the formation of the bonus fund were modified. One half of the bonus fund accruing to enterprises according to the normatives can be used only if enterprises fulfil two additional targets which reflect efficiency and the main targets of the plan. A failure to fulfil these targets entails a reduction in the size of the fund. The bonus fund increases by a share in the returns from higher prices for first quality products and in increased returns from foreign trade.[27]

From what has been said, it follows that in all three countries the centre not only sets indicators and normatives, but—what is especially characteristic of the direct system—these indicators are linked to plan targets. This is only logical since annual plans with their binding targets are the main instrument of a centralised system of management. What is no less characteristic of the direct system is that the vast part of the bonus fund is assigned to enterprises in advance by the centre. All this means that the size of the bonus fund is regulated by administrative methods.

As is well known, the imposition of binding targets encourages enterprises, *inter alia*, to conceal reserves and makes them reluctant to accept more stringent plans. This behaviour is motivated by the fear that, if they reveal their reserves by accepting demanding plans, the centre will abuse its right of control and will impose higher targets for the next year, a fear which, considering what happens in practice, is more than justifiable (see further, Chapter 6). To counter these tendencies, all the countries use more or less the same devices. Enterprises are offered higher allocations to their bonus funds for a voluntary acceptance of higher targets. On the other hand, over-fulfilment of voluntarily accepted higher targets is penalised in order to induce enterprises to accept realistic plans and thus make it easier for the planners to coordinate economic activities.

To encourage enterprises to accept demanding plans but also to extend their time horizon in decision-making, the Soviet Union is trying to use long-term normatives which have to ensure to enterprises some degree of certainty and predictability with regard to bonuses. Since 1971 the original role of normatives has been limited to the second stage of planning and implementation of plans. The GDR experimented with long-term normatives in 1969–70 and dropped them as soon as it embarked on a recentralisation of economic management.

What is also common to the incentive systems discussed is that profit as an indicator gradually loses its importance,[28] while material indicators play an increasing role. There is a tendency to slip back to the pre-reform system. This is more obvious if we compare the present incentive system with the one which came into being with the reforms of the second half of the 1960s.

We can imagine a system in which the size of the bonus fund is not assigned as a fixed proportion of a planned target, but is made dependent on the degree of increase in profit (or other indicators) over the previous year. This means that the centre does not assign a bonus fund; what it assigns to enterprises is a success indicator which measures performance (it need not necessarily be profit) and a normative which relates the increase in the bonus fund to an increase in the success indicator. Thus the size of the bonus fund is not dependent on the fulfilment of planned targets. Such a system, which is somewhere between the direct and the indirect, can be regarded as mixed. The aim of such a system is to overcome the shortcomings of the direct system that result from linking the bonus fund to centrally planned output targets—tendencies to market disequilibria and reluctance of enterprises to adopt demanding plans. In the countries under review this characterisation fits (with some qualifications) the Polish incentive system for managerial staff (which will be discussed later).

The Soviet incentive system in 1966–70 and, above all the GDR's in 1969–70, resembled the mixed system to a certain degree. In the Soviet system the bonus fund was supposed to be a varying percentage of the wage-bill, depending on two success indicators: sales (or profit) and profitability. For each percentage increase in sales, planned by enterprises and approved by the authorities, over the previous year, the bonus fund increased by a normative computed in terms of a percentage of the wage-bill. The same held true for each percentage of planned annual profitability. These normatives were supposed to be long-term.[29]

In the GDR (in 1969–70) the bonus fund consisted of two parts: basic allocation and increment. The first part was determined as a percentage share in the net profit produced in the previous year (basic normative) and the increment as a percentage share in the increment in profit (incremental normative). Both normatives were fixed in advance for two years. This incentive system meant, no doubt, a certain retreat from the old strict administrative system, but not a break with it. The East German authorities only went half way; they did not give up entirely the system of binding targets. The full allocation of funds to the bonus fund according to the produced net profit could be made, provided enterprises fulfilled the two material targets assigned to them by supervisory agencies.[30]

Let us now turn to the formation of the bonus fund in an indirect system. We have already indicated that in such a system enterprises themselves determine the amount which will be used for bonuses. In making such a decision they have, however, to consider the different constraints; they have to take into account the possible amount of funds available for bonuses, their alternative uses and the political and social consequences. The centre sets the framework in which the decision must be made; it is left to enterprises to make the choice.

Before proceeding, we must deal at least briefly with the problem of regulation of bonuses. Bonuses are part of employment incomes, and therefore planned economies are interested in their regulation for the same reasons as basic wages. In direct and mixed systems it is difficult to distinguish between formation and regulation of the bonus fund. As in the case of basic wages, these two processes coalesce; formation already includes elements of regulation. By contrast, in an indirect system these two processes are clearly separated. Formation of the bonus fund, like the wage-bill, is (or at least is to a great degree) in the jurisdiction of enterprises, whereas regulation is in the hands of the centre and is carried out by indirect methods, mainly by taxes.

The Czechoslovak (of 1966–9) and the Hungarian (mainly of 1976) systems of bonus fund formation and regulation can be regarded as indirect. This is more true of the former than the latter. The Czechoslovak reform abolished the assignment of the wage-bill by the centre and made it a function of performance. Enterprises themselves determined how much of the funds available for distribution would be used for bonuses. The growth of average wages including bonuses was regulated by taxes. In Hungary—unlike the other countries—the bonus fund depends only on the produced profit. Since 1976 enterprises themselves have had the right to divide up profit after taxation and to determine how much of it will be used for the bonus fund, within, of course, the framework of tax and other constraints.[31]

INCENTIVE SYSTEM FOR TOP MANAGERS

The incentive system for top managers cannot differ in principle from the general incentive system (i.e. for all the personnel) since both are important components of the system of management. This does not mean that the general incentive system and the incentive system for managers must be identical in all respects. It should not be forgotten that the activities of top managers are of a great variety and affect many aspects of enterprise performance and expansion. In order to control these activities more closely, the centre can find it advisable to apply a more comprehensive incentive system for top managers than for other personnel. The centre's control necessarily grows if the bonuses of top managers are determined through a subjective evaluation rather than by objective criteria. In countries with what we called a direct incentive system, incentives for managers are also under the strict control of the centre. Yet the methods applied in individual countries are not identical.

In the USSR the bonuses for top managers depend on the fulfilment and over-fulfilment of targets which are relevant to the formation and size of the incentive fund. When in 1971 the rules for the formation of the bonus fund were changed and new indicators were introduced, the incentive system for top managers underwent similar changes. The provisions of 1972 put the greatest stress on the fulfilment of productivity targets, on output of higher quality products and on new products.[32] The fundamental principles for awarding bonuses issued in 1977 underline the rules mentioned.[33] As will be shown later, the bonus fund is not the only source of bonuses.

In Czechoslovakia the so-called annual reward (the only bonus) for top managers depends on the results of the evaluation of their activities by supervisory agencies. This evaluation covers many aspects of managerial activities and the way they are reflected in annual economic results. In the quinquennium 1971–5 it included not only the extent of fulfilment of planned quantitative targets (delivery of goods abroad and to the domestic market, supply of investment goods, etc.) and qualitative targets (costs, profit, productivity), and implementation of technological plans, but also the degree to which rules regarding expenditure for wages and labour saving were observed. An important aspect in determining the size of the annual reward was the willingness of managers to accept higher targets.[34] For the period of the sixth five-year plan (1976–80) the rules have been tightened; the evaluation must in principle include fulfilment of planned targets for profit and share of costs in the value of marketed output, as well as two targets reflecting the development of science and technology. A failure to fulfil any of these targets is penalised by a reduction in the bonus of 20 per cent for each.[35]

In the GDR the year-end bonus of directors of enterprises and associations is determined by supervisory agencies. In the period 1971–4 it depended on the fulfilment of annual targets, primarily those which were assigned by the supervising organ as performance criteria.[36,37] There is no mention in the legal provisions concerning the formation and distribution of the bonus fund as to whether over-fulfilment of targets is rewarded. D. Granick argues that the motivation system for German top managers is designed in a different way from the traditional Soviet system. While in the Soviet system the manager is stimulated to over-fulfil plan targets (maximising model), the German manager is only encouraged to fulfil targets (satisficing model). Bonuses are granted '... according to subjective rather than objective evaluations of managerial performance.'[38]

As already mentioned, the Polish reform of 1973 introduced a special fund for managerial staff. This is fed from produced profit, and its size depends on the amount of produced net profit over the previous year and on a long-term normative. The size of the fund is regulated by an indirect method, taxation. (In this regard it resembles the indirect system.) The increment in the bonus fund over the previous year expressed as a percentage of the wage-bill for managers is subject to steep progressive taxation. Payment of bonuses is, however, linked to the fulfilment of two targets chosen from among the following: reduction of costs, improvement of quality of products, export deliveries, etc.[39]

The present Hungarian incentive system for managers is complicated compared with the system which the reform of 1968 brought into existence. According to that reform, the single bonus fund provided bonuses in a differentiated way for all employees including top managers. The conflicts produced by this system together with much disillusionment with profit prompted Hungarian authorities to introduce a new system in 1976. The bulk of bonuses for top managers now comes from the wage-bill but is not included in the wage-bill for tax purposes. (From the bonus fund fed by profit they receive only a year-end reward which is fixed in the same way as it is for other employees—it is proportional to basic salaries and service years.) Top managers receive an annual premium which is linked to the fulfilment of indicators, preferably to profitability (or to gross income per employee or value added, i.e. to an indicator to which wage regulation is linked in the enterprise in question). In addition managers receive an annual reward based on a comprehensive evaluation of their activities.[40] If D. Granick's classification of incentive models is used, it is possible to argue that what the Hungarians have is a combination of a maximising and satisficing model, with greater stress on the first.[41]

The changes in the Hungarian incentive system for managers mean, no doubt, a certain retreat from the principles of the indirect system. Despite this, the Hungarian system has retained important characteristics of an indirect system. The rules of the game are clearly spelled out, indicators and normatives are fixed in advance, and all these are supposed to be in force for a long time (probably for five years). What is more important is that profitability and other indicators are not assigned as binding targets, but only as yardsticks of performance.

With the exception of Czechoslovakia,[42] bonuses of top managers nowadays come from more than one source; e.g. in the USSR top managers also receive bonuses for the introduction of new technology and for production of consumer goods from waste, etc.[43] Similarly,

Polish managers also receive bonuses from the fund for technological progress, for the implementation of export tasks, etc.[44] According to D. Granick, German top managers receive bonuses from a ministerial fund.[45] In all the countries under review top managers also receive year-end rewards[46] which, with the exception of Poland, are paid from the bonus fund.

THE SHARE OF BONUSES IN TOTAL AVERAGE WAGES

Some authors believe that one of the reasons for inflationary pressures are the high bonuses paid out, mainly to managers. Unfortunately they usually do not support their statements with statistical data. Available statistical data show that bonuses in the countries under review range from 10 per cent to 20 per cent of employment incomes, if bonuses paid from the wage-bill are also included. If only bonuses from the bonus fund are considered, then their weight is much smaller—5–12 per cent, as already mentioned. It is debatable to what extent the bonuses derived from the wage-bill are genuine bonuses in the sense that they are a variable part of wages.

The availability of data for individual countries varies considerably. In most countries there are no figures available on the bonuses of top managers. In the USSR in 1973 the share of bonuses paid out from the bonus fund in industry amounted to 8.1 per cent of average wages. If bonuses from the wage-bill are included, the share was 16.3 per cent. Of course the size of bonuses is differentiated by socio-economic groups. The bonuses of manual workers which come from both sources (the wage-bill and the bonus fund) accounted for 15.2 per cent of their incomes, whereas bonuses from the bonus fund alone were 5 per cent. Engineering-technical personnel receive bonuses only from the bonus fund, and in 1973 they amounted to 22.1 per cent of their employment incomes.[47]

Table 4.1 shows that the share of bonuses in average wages substantially increased;[48] in 1973 it was (taking into consideration bonuses from both sources) 171 per cent higher than in 1960, but only 39 per cent higher than in 1967 when the reform of 1965 was already in operation. For the same period (1967–73) the bonus fund for industry as a share of the wage-bill of industry was quite stable.[49]

The figures indicated are for whole socioeconomic groups. Top managers, of course, receive relatively higher bonuses in addition to

TABLE 4.1 Bonuses in industry as percentages of total average wages in 1960–73

	1960	1965	1966	1967	1968	1969	1970	1973
All employees	6.6	8.7	10.3	11.7	12.9	14.4	15.4	16.3
Manual workers	6.0	8.0	9.2	10.2	11.4	12.9	13.8	15.2
Engineering-technical staff	11.0	13.7	16.6	19.2	20.5	21.9	23.6	22.1
Clerical staff	4.6	7.2	12.4	14.4	16.8	18.3	20.1	

SOURCE
N. Fedorenko and P. Bunicha (eds.), 1973, p. 249 and Iu. Artemov, *Voprosy ekonomiki*, no. 5, 1975.

their basic salaries. Though it is the average figures which are relevant to our study, it should be mentioned that an inquiry made by the Russian Gosbank in 1974 showed that 23.5 per cent of the top managers received bonuses from all sources equal to 38–50 per cent of their incomes; 48.8 per cent received 51–60 per cent and 16.3 per cent received 65 per cent. For the remainder, 11.4 per cent of the top managers who certainly received more than 65 per cent, no figures are indicated.[50] Recently, in the 1977 regulations for the awarding of bonuses, a maximum 50 per cent (in extraordinary cases 60 per cent) of basic salaries was made the ceiling on managers' bonuses from the incentive fund.[51] It is, however, not clear what the limits are nowadays for bonuses from other sources.[52]

In Czechoslovakia the bonus fund has played a much smaller role than in the USSR. In 1973 the bonus fund in the socialist sector amounted to 3.9 per cent of average wages.[53, 54] Figures for industry show that at the same time the bonuses from the bonus fund accounted for 1.4 per cent of manual workers' average wages and 3.5 per cent of the incomes of technical and managerial staff. If bonuses from the wage-bill are included, then the respective figures are 19.1 per cent and 20.1 per cent.[55] No figures on the annual rewards of top managers are available. What is known is that annual rewards for the top echelon of managers are fixed in the form of average limits for branches and associations. For 1976 the maximum was fixed at 25 per cent, with the exception of some branches where 35 per cent of basic salaries is payable. Top managers are allowed to receive 5 per cent above the limit.[56]

Recent Polish statistical year books only publish figures on the factory fund, which is primarily used for year-end rewards. The figures show that the fund has only increased slightly. In 1960 the fund represented 4.2 per cent of the global wage fund, whereas in 1975 it amounted to 4.6 per cent.[57] We have not come across figures on the managerial fund. What is known is that bonuses of managers before 1972 were high, ranging from 50–80 per cent of incomes. In 1972 salaries of managers were increased by introducing quite differentiated supplements to basic salaries depending on the position held in the enterprise, and bonuses of top managers were limited, according to one source, to 20 per cent of basic salaries.[58]

In Hungary the bonus fund in state industry amounted in the first year of the reform to 8.7 per cent of the wage fund, but in 1970 to only 5.5 per cent.[59] After the modifications of 1971 the bonus fund as a share of the wage fund almost doubled, and in 1975 it was almost 12 per cent.[60] Bonuses themselves did not increase at such a pace; as is known, the bonus fund was and is also used for other purposes, mostly for taxes. In 1968 the bonuses of manual workers amounted to 4.8 per cent of their basic wages and the bonuses of technicians to 11 per cent of their basic salaries, in 1971 to 5.5 per cent and 12.5 per cent respectively, and in 1973 to 6 per cent and 14 per cent respectively.[61] As already mentioned, top managers nowadays receive, apart from the year-end reward, an annual premium and annual reward as well. The annual premium can be a maximum 30 per cent of the basic salary, and the annual reward ranges between 10–20 per cent of the basic salary, depending on the size of the enterprise.[62] In 1973 bonuses of top managers constituted on the average 32 per cent of their incomes; in large enterprises the bonuses were much higher.[63]

The maximum amount per worker which can be allotted to the bonus fund in the GDR is usually set by legislation. For the year 1972 it was DM900.[64, 65] The major part of the bonus fund is used for year-end rewards.[66] The average year-end reward for 1972 was DM650 and for 1973 DM711, which corresponded to approximately one month's salary.[67] There are no figures available on Germany's top managers' bonuses.[68]

The reforms of the 1960s brought about an increase in bonuses, in particular in the USSR. The increased size of bonuses boosted, of course, total average wages, but not dramatically. In the USSR, after a rapid growth in the 1960s, the growth of bonuses in the seventies as a share of total average wages slowed down remarkably (see Table 4.1).

The growth of the bonus fund is nowadays kept under strict control. In the USSR and the GDR, apart from assigning the bonus fund to enterprises as an absolute sum (planned size), the planners also put a ceiling on its actual size. In Czechoslovakia there is no ceiling, but the bonus fund plays a very small role. Poland (in the case of the managerial bonus fund) and Hungary exercise control through heavy progressive taxation. For all these reasons, the assertion that the incentive funds are a source of inflation is not borne out by the present practice.

5 Wage Control and Success Indicators

In the preceding two chapters the regulation of basic wages and bonuses has been discussed on a very general level. Attention has been primarily devoted to systemic aspects, to the way the regulation is carried out, whether directly or indirectly, and partially to its impact on price stability. The choice of indicators, the systemic significance of individual indicators and the reasons for their choice have not been touched on until now. This is the object of the present chapter.

In a centrally planned economy an evaluation indicator (or indicators) must exist, though its role varies according to the nature of the system. In a centralised system, as already noted, the success indicator is used to measure the fulfilment of plan targets, whereas in a decentralised system it is a yardstick of the actual performance. The evaluation indicator may serve as a regulator of both wages and bonuses or of either one. It is obvious that in a system with an integrated wage-bill, only one indicator is needed. Thus in Czechoslovakia, for example, gross income, the indicator adopted for the evaluation of performance, served in 1966–9 for the regulation of wages as well as bonuses. Nowadays, as already mentioned, in all the countries under examination enterprises have bonus funds separate from the wage-bills; therefore they are regulated by two different indicators.

In the second half of the 1960s all the countries which undertook a reform were engaged in a debate—openly or behind the scenes—about the advantages and disadvantages of different indicators for measuring economic activities and for motivating enterprises to act in accordance with the objectives of national plans. In general the tendency was to replace gross output indicators with a synthetic indicator, or at least to let a synthetic indicator play a significant role. Some countries considered the selection of the evaluation indicator from the point of view of its suitability for the regulation of both wages and bonuses. This was the case in Czechoslovakia and Hungary (which soon opted for a separate indicator for wage regulation). Since the USSR and the GDR did not intend to make changes in their SWR and were determined to use the bonus fund as the main incentive, they were only looking for success indicators which could be applied to the regulation of the incentive fund. Poland, with its reform of 1973, opted from the beginning for making the wage growth regulator the main evaluation indicator.

The question may be raised: was the selection of evaluation indicators influenced by systemic considerations? It seems that the answer can only be affirmative. Decentralised systems have a tendency to rely on one synthetic evaluation indicator, whereas centralised systems prefer to depend on several indicators, on a combination of quantitative and qualitative (or synthetic) indicators. It is known that the USSR in 1965 resorted to sales and profitability. After 1971 profit was downgraded though not dropped, and additional quantitative indicators were introduced. East Germany, even when adopting profit as the main indicator in the 1960s, never entirely dropped quantitative indicators. As will be shown later, there are also differences between indirect and direct systems of wage regulation in the nature of the indicators used. First, a brief survey of opinions in the Soviet bloc countries about the advantages and disadvantages of the main evaluation indicators adopted during the reforms of the 1960s will be presented. This will give some insight into the reasons which were decisive for the choice of indicators. Only after this survey will we turn to a discussion of indicators for wage regulation.

PROFIT VERSUS GROSS INCOME—SURVEY OF OPINIONS

In a decentralised system the choice is between profit and gross income (and their modifications) simply because they are the most synthetic indicators. In all the countries under examination, with the exception of the USSR, pre-reform debates revolved around the choice between these two indicators.

As already mentioned, the Czechoslovak reform (1966–9) adopted gross income as an indicator of performance, regulator of wage growth (including bonuses) and also as a basis for taxation. Many economists disapproved of the decision; particularly in 1967–8, when a more liberal atmosphere came into being, a debate developed. M. Sokol, who was one of the architects of the reform, expressed perhaps the most articulately the ideas of the reformers in this regard. According to Sokol, at the heart of the problem of selecting an indicator is the question as to which is the most adequate instrument for promoting the material interest of enterprises and their personnel. He believes that this could be best achieved by a series of taxes levied on the evaluation indicator and also on the wage-bill. These taxes are supposed to serve as a criterion of efficiency. On the one hand, they have to show enterprises to what extent their performance is efficient, and, on the other hand, they must exert pressure on enterprises to increase the efficiency of their operations.[1]

Sokol thus departs from the idea that taxation should be based on the ability to pay, as this only makes sense—to him—if the purpose of taxation is to mitigate income differences. In that case it is correct to levy taxes according to the principle that the more you earn, the more you pay. However if the aim of taxation is to enforce and improve efficiency, then a different principle must govern it—the more you earn, the more you have for remuneration.[2] It is obvious that the purpose of Sokol's proposition was to terminate the old practice which, in fact, rewarded enterprises having a poor performance and penalised enterprises which performed well. Badly performing enterprises usually got 'softer' plan targets, whereas enterprises with a good performance were exposed to taut output targets and higher taxes. He believes that the application of taxation as a criterion of efficiency is essential if structural changes aimed at making the economy more efficient are to be achieved. To make his advocated taxation principle workable, Sokol had earlier called for the application of a proportional tax on revenues of all enterprises,[3] and the reform did introduce a uniform tax rate on gross income for industrial enterprises.

Sokol believes that profit is not a suitable basis for his principle of taxation under socialism. Profit does not play the same role in a socialist economy as in a capitalist. Profits made by capitalists form their income and are used for two purposes: for their private consumption and for investment. The tax on profit under capitalism is in substance a genuine income tax. Due to the way profits are used, there is a long-term tendency to tax more heavily the part of profit used for private consumption than the part used for investment and technological development.

Under socialism the profit of an enterprise is not the income of its personnel to be used for consumption; it is earmarked primarily for investment. Therefore wages should be the main target of taxation 'which is in line with the fact that in our conditions there is (in enterprises) a greater interest in wages than profit.'[4] For this reason Sokol suggests making gross income, which contains wages (and profit too) the basis for taxation.

According to Z. Kodet, in the final analysis only one important aspect was decisive in the selection of gross income, and that was the desire to create conditions for an economic regulation of wages. Like Sokol,[5] he believes that the use of profit as an indicator and base of taxation would not allow the direct assigning of the wage-bill from the centre to be abandoned. Kodet does not explain why, but he must have had in mind that such a system would not force enterprises to use objective criteria for wage formation. If managers of enterprises are entitled to make decisions about wages, they may be tempted to pay excessive wages and thus reduce the amount of profit. Such a step may lead not only to unwanted (from the viewpoint of the society) wage increases, but in addition will reduce state revenues due to lower profits. By contrast, taxation of gross income (which means simultaneously taxation on the portion earmarked for wages) enables the introduction of an economic regulation of wages, according to the author.[6] In addition the adoption of gross income makes possible a regulation of both wages and bonuses by a single evaluation indicator.[7] In practice this means that enterprises are given the right to determine—of course, in the framework of legal regulations—how much of the funds at their disposal will be used for average wage increases and how much for bonuses.

Kosta and Levcik argue that promoting profit to an indicator would necessarily mean taxing profit as in a capitalist economy. Under capitalism taxation of profit is acceptable since capitalists are interested in its maximisation and protect it from the working class. By contrast, under socialism there is no class which is dependent on profit due to its position in production and which would defend it against the workers.[8]

What these authors only touched on was expressed in a much more articulate way in the recent Polish debate in connection with the reform of 1973 which adopted output added as an indicator. J. Pajestka, one of the leading architects of the Polish reform, argued against profit by stating that under socialism the managers are not a sufficiently strong social power to be able to resist pressure from the personnel for wage increases. Therefore if profit is used as an indicator, the central authority must directly assign binding targets for profit.[9] In other words if the assignment of profit from the centre is to be avoided, profit should not be used as an indicator. Pajestka in a way agrees with Sokol and Kodet that profit requires a central assignment of a binding target. However, he had in mind that profit must be the binding target, whereas the other two were thinking of the wage-bill.

There is a further important argument against profit—it may cause conflict of interest between management and personnel.[10] The conflict may have its origin in the so-called managerial system (based on the one-man management principle) but can also be found in a self-management system. It can be assumed that in a managerial system managers are intensely interested in maximising profits, and therefore they may try to achieve this at the expense of wages, thus giving birth to conflicts.[11] This is mainly true if managers get a higher share in bonuses than other personnel, which is the usual case.

In a self-management system the developing conflict may be the other way round. If workers' councils make decisions concerning the distribution of income of enterprises, they simultaneously influence the amount of profit on which the bonuses of managers depend. If they take the opportunity to maximise wage increases, the amount of profit may be smaller than in other enterprises with the same conditions but with less efficient managers. This conflict of interest may weaken managerial interest in the economic development of enterprises. On the other hand, if the incentive is gross income the managerial interest would not be affected by a maximisation of wages. This is one of the arguments that some Czechoslovak reformers put forward in favour of gross income.[12]

The opponents of profit—as has been shown—used mostly institutional arguments. Czechoslovak economists regarded the extension of the economic autonomy of enterprises as the primary goal of the economic reform. They believed that greater economic autonomy combined with elements of the market mechanism would also act in the direction of greater efficiency. Certain anti-reformists also objected to profit for institutional reasons. Some of the opponents of E. Liberman argued along these lines. For example A. Bakhurin and A. Pervukhin objected to the promotion of profit as the only indicator by contending that profit is not the main purpose of socialist production.[13] A. Zverev rejects profitability also because its definition (profit over invested capital) means an admission that '. . . profit is created not only by the workers' labour but also by the fixed and current assets.'[14]

In the arguments of those who preferred profit to gross income, institutional reasoning also played an important role. Gross income was at that time associated with the Yugoslav self-management system. Some planners and politicians objected to gross income just because they were against self-management or were afraid to be associated with it. 'If more thoroughly analysed'—writes B. Sulyok—'the advantages and disadvantages (of gross income) prove to be associated with the concept of the "autonomous enterprise" being the object of group-property and this concept was discarded when the basic ideas of our reform have been established.'[15] In this connection it should be stressed that one of the reasons why Czechoslovakia opted for gross income was just the conviction of many reformers that the Yugoslav system had proved itself in many respects, so that some of its elements should be adopted. Many reformers were particularly impressed by the Yugoslav system of workers' councils.[16]

Another great objection to gross income was that the same work would be differently rewarded. B. Csikós-Nagy, one of the architects of the new Hungarian economic system, characterised gross income and profit in the following way: in the case of gross income '... wages are proportionate to accomplishment ... while the same work is of necessity differently rewarded in various enterprises'; if profit is the incentive, 'wages paid for the total product are independent of the total value of the commodities produced ... In the case of this type of incentive, a deterioration in economic efficiency does not reduce the gross income but the profits, and may even put enterprises in the red.'[17] Even if he does not label gross income explicitly as an unjust indicator and prone to inefficiency, the context in which the characterisation of profit is made, however, leaves such an impression. See also p. 99.

Two other economists argued more overtly that a consistent application of gross income 'would make wages entirely dependent on gross income, without the guarantee of a minimum wage, although the individual worker is not responsible for the entire activity of the enterprise as a whole, nor for its gross income'.[18] Some inferred from the application of gross income that there was a danger of generating an excessive differentiation of wages which might lead to inflationary pressures.[19]

All these views must have been based on the assumption that wage growth should not be linked to profit (otherwise wage differentiation would be even higher) and that some special arrangement should be made for wage determination. Many preferred to let enterprises pay approximately the same reward for the same work;[20] in other words, they wanted to break off the linkage between wage growth and the performance of enterprises. As will be shown later, the final design of the reform of 1968 did not encompass this idea.

Some Czechoslovak[21] and East German economists also opposed gross income for institutional reasons. H. Nick in East Germany argued that in contrast to capitalism where the main goal of production is profit, under socialism it is the physical volume of national income per capita, and therefore this volume is the yardstick of performance on the macroeconomic level. Profit in turn should measure the contribution of enterprises to national income.[22] He dismissed possible charges of inconsistency by referring to the circumstance that wages—one of the two main components of gross income—are regulated by the centre so that there is no sense including them in the indicator of performance of enterprises. In addition an adoption of gross income would necessarily have ownership repercussions.[23]

In this connection it is worthwhile mentioning that the Soviet economist, V. Sitnin, when arguing recently in favour of net output, also started out with the contention that net output (which is similar to gross income) is the goal of socialist production. He arrived, however, at a different conclusion from H. Nick; to him net output should be used as an indicator on all levels of the economy for the sake of a uniform approach.[24] Seifert, Pohl and Maier are even more specific than Nick. They warn that the adoption of gross income as an indicator would mean that decision-making about the proportion between consumption and accumulation would shift to enterprises to a great degree.[25] Hungarian opponents of profit favoured gross income precisely because they believed that enterprises should be given the right to distribute gross income.[26]

The debate on profit versus gross income was not confined to institutional arguments. It was also conducted on the efficiency plane. The most important argument of the adherents of profit was that profit is the most synthetic (reflects the effectiveness of all the special indicators) and objective indicator of performance, and that it (or more precisely, profitability) encourages an efficient utilisation of fixed and working capital, and stimulates labour saving and optimum size of inventories, etc.[27] In the 1960s those Soviet economists who took a position in favour of profit also used efficiency arguments. At that time gross income was not given any thought; to what degree this was due to ideological considerations can only be guessed. Only in the 1970s when profit and profitability were downgraded as indicators did considerable interest in net output emerge (see further Chapter 7). But again net output is considered in comparison with gross value of output as profit was in the past.

In his well-known article in *Pravda* which introduced the debate on the possible design of the economic reform, E. Liberman forcefully argued in favour of the introduction of profitability as the main evaluation indicator. His main argument was that this would make enterprises more interested in taut plans, cost reductions and utilisation of invested capital.[28] He, together with V. Nemchinov (who argued much along the same lines),[29] wanted to make profitability an evaluation indicator which would be planned by enterprises themselves and also to introduce long-term normatives which would determine the size of the bonus fund depending on the level of achieved profitability. Liberman explicitly suggested that enterprises should only be assigned targets by the centre for volume of output, product mix and delivery schedules.

Profit was also indirectly defended through inefficiency effects being attributed to gross income. B. Csikós-Nagy, referring to B. Ward's[30] well-known paper, writes: 'In the case of the profit motive, production in response to the market mechanism follows the changes of demand-supply relations and price changes in a rational course but in the case of the income motive this automatism works irrationally. This is the point, among others, why Hungary insisted on the profit motive...'[31] Two other Hungarian economists argued that gross income would orient enterprises one-sidedly to a saving '... of embodied labour and would fail to promote the optimum utilisation of live labour. This would lead to an exaggerated increase of personal incomes at the cost of economic growth. Nor would this indicator secure for the state budget its proper share of the increment of enterprise incomes.'[32]

The defect of gross income in not encouraging optimum utilisation of labour can be partially cured by applying it, not as an absolute sum, but per employee. In this connection it is worthwhile mentioning that Hungarian developments in 1968–70 did not prove the authors' argument. In 1971 Hungary had to change over from profit to gross income per employee as a regulator of wage growth just because of excessive hoarding of labour (see p. 154).

Some argue that gross income encourages enterprises to shifts to more profitable products and, therefore, to an insufficient regard for consumer demand. Under conditions of a seller's market this is a relevant statement with the qualification that it can also be levelled against profit, a qualification which many economists actually make.

From what has been said about the debates, it is clear that the controversy, gross income versus profit, was not, and after all could not be, free of ideological considerations. If one regards the self-management system with all its consequences as an ideal, he would rather favour gross income; and, vice versa, opponents of self-management would be more likely to prefer profit. However this is not an absolute rule. The Poles adopted output added (which is similar to gross income) for wage-bill regulation without introducing a self-management system.

Another problem—which is at the heart of the controversy for those economists who want at least some kind of decentralisation—is which of the two indicators enables an economic regulation of wages without exposing the economy to upward pressure on wages. Here, the adherents of gross income have a good point against profit, but it is not as strong as they would like to make people believe. It is quite conceivable to have an incentive system based on profit and yet have at the same time an indirect regulation of wages. It is possible to forestall undesired wage increases, which an incentive system based on profit may encourage, by a properly designed system of taxation. A tax on gross income is simultaneously a tax on wages; however, it can only affect wages at the same rate as profit. What about taxing profit and wages separately?[33] This would enable the imposition of a progressive tax on wages which might lessen the pressure for wage increases.

Adherents of gross income have also a strong point in their contention that profit may give rise to conflict of interest. However, even here some qualifications may be raised. In a managerial system it is possible to avert conflicts of interest if the material interest of the whole work force is linked to profit,[34] and bonuses have approximately the same weight in the pay of manual and non-manual workers, including managers. If, however, much higher bonuses are given to managers for incentive reasons, a conflict of interest may develop.

No doubt profit is a more objective evaluation indicator of enterprises' performance than gross income. This is not so much true of profit as an absolute sum (amount of profit) as of profit as a relative sum (profitability computed as profit over fixed and working capital). The amount of profit can be increased without a better utilisation of factors of production, simply by expanding output (hiring more labour). This is not to say that profitability cannot be increased without an increase in efficiency; there are several ways to manipulate it. In addition it is not advantageous from the viewpoint of efficiency to use profitability in all sectors of the economy.[35]

Without intending to underestimate the advantages of profit as an indicator of performance, we feel that systemic solutions are of greater importance, namely, the way the wage-bill and bonus fund are linked to indicators. It is known that linking wage or bonus to the fulfilment of binding targets raises inefficiencies. Therefore even the adoption of profit is not a guarantee that the system of management will become more effective than in another economy with gross income, but without binding targets. On the other hand, the replacement output targets by profit in the same system promises some improvements in efficiency. In addition the effectiveness of wage and bonus regulation depends to a great degree on the rationality of the price system. Distortions in relative prices characteristic of the Soviet bloc countries (including Hungary, though to a lesser degree)[36] reduce the objectivity of indicators (including profit) as a yardstick for the measurement of performance of different enterprises. Needless to say, such a situation reduces the effectiveness of indicators as a stimulus.

It has already been noted that all the countries nowadays use profit as *the* or one of the indicators for the incentive fund regulation. (In Poland, it is only used for the regulation of the managerial incentive fund). All the countries use amount of profit; the USSR also uses profitability but both (profit and profitability) are used in a very reduced role compared to the original intentions of the reform of 1965. Though profitability—as has been mentioned—is a more synthetic and objective indicator, there is a tendency to adopt amount of profit rather than profitability. Profit is easier to administer, to assign targets in and to check for fulfilment or over-fulfilment. Calculation of profitability (as profit over fixed and working capital) also requires determining the value of fixed and working capital and evaluating frequent changes in the structure of fixed capital. In the discussion about the intended reform of 1970, many Polish economists argued against profitability merely by alluding to the accounting difficulties connected with its application and operation.[37] The preference for amount of profit as indicator may also have something to do with the fact that profitability is a more rigorous indicator and, for that reason, is disliked by enterprises. The Soviet reformers argued mostly in favour of profitability. E. Liberman contends that profit alone is not a sufficient indicator of enterprises' ability to operate profitably 'since this ability depends on the quality of the labour applied'. And this is reflected in profitability by comparing profit with productive assets.[38] The reform of 1965 attributed a greater role to profitability than to profit.

It would be interesting to know why the Hungarians resorted to profit instead of profitability. To our knowledge, very little has been published on this problem. The only paper we have come across where the problem is discussed is the one already mentioned by B. Csikós-Nagy in the *New Hungarian Quarterly*. Because of its briefness, the arguments are not very clear. It seems that opting for profit was mainly motivated by the adopted price formula. It is not unlikely that unfavourable experience with profitability as a regulator of bonus fund in the period 1957–64 also contributed to the adoption of amount of profit.

INDICATORS FOR WAGE CONTROL

Up to this point we have discussed evaluation indicators in general, mostly stressing their function as regulator of the bonus fund. Let us now turn specifically to indicators for the regulation of wage growth. Their selection is not a simple task if a coherent SWR is to be the result. Many factors must be taken into consideration when making the decision.

1. Naturally, systemic aspects are the first to be considered. A decentralised system of management, combined with an indirect system of wage regulation, is incompatible with gross output indicators characteristic of the direct system. If the SWR is to be really effective, the regulator must be a net indicator[39] designed as a genuine sales indicator. In other words, the indicator must encompass both properties; one may be of greater importance than the other in different situations. This is not to say that countries with a direct system must necessarily stick to a gross indicator, though this has been the case. While countries with non-direct systems have favoured a net indicator (gross income or value added), direct systems have tended to use gross indicators (gross value of output or commodity production or sales including costs of materials). There are no valid systemic reasons why a direct system could not opt for a net indicator. (On the other hand, it is questionable whether for systemic reasons it can afford to make the net output indicator a genuine sales indicator.) After all—as will be shown later—the Soviet Union some time ago, and Czechoslovakia recently, embarked on experiments with net output. But what is optional in a direct system is a necessity for an indirect system. The adoption of a net indicator which is at the same time a sales indicator does not turn a system into an indirect one, but an indirect system cannot properly function without it. There are several reasons for this:

(a) In a decentralised system authorities do not impose—as a matter of principle—binding targets on enterprises as is the case in a centralised system. Authorities rely on market forces to induce enterprises to produce goods which are in demand. And a sales indicator is important to make the market forces work.[40] In practice the adoption of a net indicator designed as a genuine sales indicator is complicated by the fact that in many enterprises, due to their length of production cycle, annual performance cannot be expressed accurately by such an indicator. Therefore exceptions must be made. It is known, however, that the inclusion of work in progress in the indicator lends itself to manipulation.

(b) To make the system more effective it is important to reduce the possibilities of juggling with the indicator. Since a net indicator does not include material costs it can be less manipulated than a gross indicator. However, it is vulnerable to shifts in output-mix to profitable products, an activity which can be the more practised the greater the sellers' market. Under conditions of net indicators, the weight of such shifts is much bigger (due to the greater share of profit in the indicator) than in the case of gross indicators. The sales property of the net indicator may dampen such activities.

(c) Last but not least, the sales indicator has to prevent enterprises producing unsaleable goods for the sake of maximising the wage-bill. As will be shown later, countries with an administrative system try to tackle this problem in different ways. At the beginning of the 1960s the Polish planners—to mention one example—in an effort to discourage production of unsaleable goods, made over-fulfilment of output targets by enterprises contingent on special financing and approval by supervisory bodies.[41] In an indirect system the sales indicator is called on to fulfil the function which is fulfiled by administrative methods in a direct system.

2. The selection of a wage-bill regulator must be coordinated with 'what is regulated'. This is even more important in a system where regulation of wages is exercised by indirect methods. If for example a country with an indirect system regards as its first priority the prevention of inflationary pressures, it will probably opt for average wage regulation. In such a case, it would be advisable to combine it with a productivity indicator in whose formula the numerator is, of course, a net indicator designed as a sales indicator. The rationale of such an arrangement is to discourage excessive growth of employment, which might be stimulated by average wage regulation and which might be a source of inflationary pressure, thus contravening the original objec-

tive of selecting average wage regulation. True, the linkage of wage growth to an absolute indicator encompasses a built-in stimulus to productivity increases. However, for various reasons it may not be strong enough so that the productivity indicator must strengthen this stimulus.

Whether the combination of average wage or wage-bill regulation with a certain indicator will work well depends on many factors—one important one being the extent to which the planners are able to make the indicator an objective reflection of performance. The realisation of this principle requires not only the selection of the most objective indicator possible but also a definition of performance which would encompass only individual enterprises' own efforts.[42]

3. It is obvious that the systems of wage regulation and bonus fund regulation must be complementary and not conflicting if they are to be effective. The same is true of the wage growth regulator and the bonus fund regulator. Wages are a reward for labour input, whereas bonuses coming from the bonus fund are primarily a share in the returns on economic activities. On the basis of this statement one could argue that the size of the bonus fund should be linked to profit which best reflects the returns on economic activities, while the wage-bill growth should be directly linked to labour intensity of production. This first suggestion has been wholly or partially adopted in the countries of the Soviet bloc. The implementation of the second suggestion would, however, require the calculation of work norms for all work to be done—an activity which would be very costly and, because it would have to be done in most cases by enterprises themselves, far from accurate[43]—and, therefore, would hinder technological progress and efforts directed to a better organisation of production. What is no less important, it would stimulate claims for a higher and higher wage-bill, which would reduce the amount of profit and thus its effectiveness as a regulator of the bonus fund. On the other hand, an indicator which contains in addition to wages the magnitude which reflects the returns on economic activities—this means profit—may reduce threats to technological progress and thus push in the same direction as profit as a regulator of the bonus fund. And this is also the reason why wage growth is not linked in any system directly to labour intensity of production but instead to a net or gross indicator, depending on the nature of the SWR.

Before discussing wage growth regulators in more concrete form, it should be stressed that up to now no country has yet linked the formation of the wage-bill to profit. Only Hungary experimented for some time (1968–70) with financing increases in basic wages (in addition to bonuses) from the incentive fund, which is fed from profit. Of course, it is also possible to use profit as an indicator for wage growth without requiring enterprises to finance wage increases from it. To our knowledge, no such experiment has yet been undertaken.[44]

There is little literature on the reasons for profit not being used as an indicator for wage growth. V. Sitnin, a Soviet author, argues that profit arises after production costs, including wages, have already been paid, so that it would be impossible (he does not explain why) to plan the wage-bill on the basis of profit or profitability.[45] It seems that there are other reasons too which may be of great importance. To begin with, to link wage increases to profit means changing basic average wages or the wage-bill (depending on what is regulated) according to fluctuations in profits. And it is known that these fluctuations are relatively large, and, what is worse, their extent is difficult to predict with accuracy. This makes the application of such a linkage very difficult in practice. Workers are reconciled to fluctuations in bonuses, but they would react with bitterness if their basic wages declined because of a drop in profits.

If wage increases are financed from profit, additional problems may arise. To calculate production costs correctly, increases in basic wages should be included in labour costs. However if they are financed from profit, they are, by definition, excluded from them. And finally, linkage of wages to profit may lead to unwanted inter-enterprise wage differentials. J. Wilczek argues that this is one of the factors which hampers a long-term linkage of wage increases to profit.[46]

The Soviet economic reform of 1965 has not brought about any significant changes in wage regulation. Though gross value of output as an evaluation indicator of plan fulfilment was replaced by other indicators, it (or in some cases, commodity production) remained as an indicator for wage growth as well as the numerator in the productivity formula at the level of enterprises. However, in various periods experiments have been made with some modified indicators.[47] Of great importance are the experiments with net output which will be discussed in Chapter 7. After Dubček's fall, the Czechoslovak decentralised system was gradually dismantled and with it also the system of wage regulation. Gross income as an indicator of performance and regulator of wage growth was replaced by marketed output. In the GDR the reform gave rise to some experiments linking wage increases to productivity, but from the information available, it appears that East Germany has not left the framework of the direct system.[48] The difference between Soviet, the present Czechoslovak, or the German indicators is really not of great importance. All are based on gross value of output, directly or indirectly.

Hungary as well as Poland (and, of course, Czechoslovakia in 1966–9) use a net indicator for wage growth regulation. Not only is there a systemic difference in the way the indicator is used—this has already been explained in Chapter 3—but also a difference in the way the net indicator is conceived. In contrast to Czechoslovakia where gross income was not designed as a consistent sales indicator, in Poland and Hungary the net indicator is at the same time a sales indicator. In addition Czechoslovakia used and Poland uses it in an absolute form, whereas Hungary applies it in the form of a productivity indicator for some sectors of the economy. Hungary is thus the only country which uses a productivity indicator for regulating growth of wages.[49] As will be shown in Chapter 9, its operational scope was narrowed in 1976, though it is still an important indicator.

Judging by the importance attached to the achievement of targets planned for the relationship between wage and productivity growth by the countries under review, one would expect that a productivity indicator would be heavily used as a regulator of wage growth. But this is not the case. Not that the productivity indicator does not play any role in the control of wage growth, but rather it is used as a supplementary device or in cases where the main regulator shows signs of failing. In Czechoslovakia and the GDR, fulfilment of the productivity target is a precondition for enterprises being allowed to use the whole planned wage-bill. (Yet in Czechoslovakia fulfilment of the productivity target is not statutory.) In Poland enterprises which have not yet been converted to the new system of planning and financing must fulfil obligatory productivity targets. Recently in the USSR additional allocation of funds to the wage-bill for over-fulfilment of output targets has been linked to productivity. The Soviet Union also encourages the fulfilment of productivity targets by making the size of the bonus fund dependent on the fulfilment of this indicator.[50] In Czechoslovakia (1967–8) the productivity indicator was used, and is probably used in Poland at the present time, for enterprises converted to the new system, because the main indicator has turned out not to be strong enough to avoid a wage drift.

Why is the productivity indicator not used as the main regulator of wage growth? There appear to be several reasons for this. First, all the SWRs have, to some extent, built-in productivity incentives. For example in the Polish system adopted in 1973 the wage-bill increase is contingent on an increase in output added. It is obvious that for a given increase in output added, the smaller the employment increase, the greater the wage increase.

Secondly, experience shows that it is not an easy task to find an objective criterion for the computation of productivity at the enterprise level, particularly if it is to be applied to the vast majority of enterprises. The present formula for productivity computation used in most of the countries (gross value of output over number of employed) is far from being objective. In Chapter 7 the shortcomings of gross value of output will be discussed. Here it need only be mentioned that this indicator can easily be tampered with by changing the product mix. D. Karpukhin and V. Rozhkova contend that a recent inquiry has shown that in many enterprises 60–85 per cent of productivity targets were achieved by shifts in product mix.[51]

Finally, the great stress on expansion of output may have made planners reluctant to link wage growth to productivity growth on the enterprise level. However, increasing shortages of labour may eventually bring about a change. Recently the Soviet Union (the GDR also) has increasingly experimented with linking wage growth to productivity growth (see Chapter 7).

6 Uniformity Versus Differentiation

In designing the SWR the planners are faced with deciding whether to apply the system uniformly throughout the economy or whether to differentiate it, taking into consideration special conditions in individual sectors, industrial branches or even enterprises. The problem of uniformity versus differentiation touches many facets of wage (and bonus) regulation—methods of linking wages and bonuses to indicators, the choice of indicators, taxation and normatives. Up to this point the role of taxation and normatives in wage control has only been briefly discussed. The topic of this chapter will also serve as a framework for the examination of that aspect.

First of all, the term 'differentiation' must be explained; the term 'uniformity' is quite straightforward. Differentiation within a uniform system must be distinguished from differentiation which goes beyond the system. An example will make clear what we have in mind. We can imagine that in some branches planned output is the indicator, in others planned sales and in still others planned profits. In other words, in all branches wage (or bonuses) growth is linked to an indicator expressed in plan targets. This is a differentiation within a uniform system (linkage to plan targets). On the other hand, we can imagine that in one branch wages are linked to a plan target and in other branches to an increase in performance over the previous year. This is a differentiation that goes beyond the system. We will deal primarily with differentiation within the system of wage regulation.

Generally, many considerations favour uniformity, and a few will be mentioned here. First of all, it is usually simpler to design a uniform system. This is particularly true when it comes to determining normatives or tax structures (in cases where wages or bonuses are regulated by taxes). The design of a uniform system does not require such complicated assessments and computations as a differentiated system. It is, however, in the nature of the uniform system that it *cannot* be equally just to all enterprises, that it benefits some and handicaps others. In addition a uniform system does not require continuous intervention by central authorities. The centre must, of course, see to it that the rules of the game are correctly formulated and adequately changed if, in the course of their implementation, it becomes clear that conceptual errors have crept into the design of the rules or that the assumptions on which they were based have changed. Once this has been accomplished, the centre's duty is to see to it that the rules are heeded by all. Finally, it can be argued that a well-thought-out system applied uniformly has the advantage of benefiting enterprises which perform well and exerts pressure on lagging enterprises to improve their performance.

However, conditions in individual sectors, branches and enterprises differ in many respects. Aside from the technological level, individual sectors differ as to the character of production (agriculture, industry). But even within industry (in the classification used in the East) great differences exist, mainly between mining and manufacturing. In the former, productivity depends very much on natural conditions, whereas in the latter this is only true to a negligible degree. With some simplification it can be contended that the products of mining are not differentiated, whereas in manufacturing a high degree of product differentiation exists. For this reason the measurement of productivity in the former can be much more objective than in the latter. In mining, the measurement can mostly be made in physical units, whereas in manufacturing, value units must be used for this purpose. The achieved level of technology on the world scale predetermines to a great degree the structure of costs of production and makes some economic units more capital intensive than others. This is to some extent modified by the priorities of socialist countries.

In addition not all units have the same possibilities for expansion for reasons which may be found in the units themselves or because of planners' priorities. Some units are supposed to grow fast, others slowly; some are even supposed to stagnate. The role they play or may play in foreign trade has, of course, an important effect on their growth. In this context, the availability of labour is of importance. In some units fast expansion is supposed to be achieved by expanding employment; in others the stress may be on increases in productivity. Last but not least, in some units the stress is on the availability of goods and services and their quality rather than on efficiency and profit.

All these differences are also due to factors which are beyond the control of enterprises; consequently a uniform system, as already mentioned, puts some units in a more favourable position than others. Therefore one could argue that a uniform system throughout the economy is not fair and may for this reason be less effective. A differentiated system, however, opens the door to bureaucratic arbitrariness from the centre. There is reason to doubt whether the centre can have sufficiently reliable data to make correct differentiations in the control provisions. On the other hand, such a system encourages enterprises to press the centre for preferential or differential treatment. The fact that government agencies are allowed to vary their regulations is a great motivation for enterprises to seek concessions.

Up to this point we have disregarded the fact that the system of management itself tends to imply a certain relation between uniformity and differentiation. Generally it can be argued that the more the system sticks to control, the less conducive it is to uniformity, and vice versa. Control as it is applied in an administrative system can be effective if the centre can differentiate the operating rules of the system according to the conditions of the economic units and if it can interfere whenever the behaviour of the units deviates from what is expected. Yet uniformity means grating enterprises certain room for manoeuvring, which necessarily means in practice that the probability of enterprises behaving according to the desired pattern is reduced. What is no less important, it makes controls more difficult since enterprises may try to justify their behaviour by referring to their special conditions. This is not to say that a direct SWR must be differentiated in all respects. Apart from important aspects which are characteristic of the system and therefore should be uniform (linkage of wage growth to fulfilment of binding plan targets), the concrete design of the system may also include some uniformity.

A decentralised system by definition means greater economic autonomy for enterprises. Frequent interference with the enterprises' economic activity undermines this autonomy, which is extended in the expectation that the enterprises will use the room given them for taking initiative for the sake of promoting efficiency. Setting the same rules of the game for all enterprises and rewarding those which perform well and penalising those which perform poorly is one of the most important tools used to achieve these goals. An important qualification must be made to this statement: uniformity can be helpful in the direction mentioned, provided differences in the starting conditions of the 'players' are not too distinct. In addition a socialist system which puts full employment at the top of its list of priorities cannot disregard the effects a uniform system may have on lagging enterprises, all the more because their conditions may be the result of consequences beyond their control.

Every system of wage (and bonus) regulation is supposed to fulfil many goals and to respond to many constraints. As already noted, it is not only supposed to protect the economy against inflation by keeping the growth of wages (and bonuses) within the limits of the growth of the consumption fund, but it also has to induce enterprises to utilise more rationally the factors of production. Also it is not supposed to upset a wanted structure of wages. All these requirements need not necessarily be complementary; some may be in conflict.

All these factors, as historical experience shows, have contributed to the rise of a situation in which the indirect SWR is not uniform throughout the economy. Thus the Czechoslovak reform (1966–9) used two systems of wage regulation. One, the dominant, was a pure indirect system; the second, limited to a fraction of enterprises, was to a great degree a carry-over from the old system. Hungary also has two systems nowadays. Apart from the dominant system, which is characterised by a linkage of wage growth to improvement in performance over the previous year, average wages or the wage-bill are in some sectors or industrial branches directly assigned to enterprises (for more see Chapters 9 and 10). Compared to 1968 or even 1971, the Hungarian system of wage regulation is today much more differentiated, more adjusted to concrete conditions. However, in several important respects (e.g., tax rate on profit and bonus fund) uniformity has been retained.

UNIFORM OR DIFFERENTIATED TAXATION?

It has already been shown that in the indirect system the regulation of wages is done by indirect methods, that is, by taxation. This was true in Czechoslovakia and is the case in Hungary (and in Poland as regards the bonus fund for managerial staff). Taxation of wages for the purpose of their regulation should be distinguished from taxation for other purposes. Taxation is used nowadays in some countries for the purpose of making labour more expensive. In all the countries of the Soviet bloc labour is inexpensive compared to other factors of production; therefore enterprises are not very interested in substitution of capital for labour. This is one of the reasons for hoarding labour in some units at a time when it is becoming increasingly scarce and when some sectors suffer from labour shortages. In order to induce enterprises to economise on labour some countries have introduced a single tax on wages paid out. Thus in Hungary enterprises pay an 8 per cent tax[1] and in Poland[2] 20 per cent on the wage-bill from produced profit. In Czechoslovakia up to 1978 there was a progressive tax on the wage-bill payable for increases in the growth of average wages over the previous year.[3] In some of the literature it was maintained that the tax served as an instrument of wage control.[4] In our opinion, it was rather a vestige of the reform of 1966–9, and its real purpose was not wage control primarily. The best proof of this is the fact that the legal provisions of 1976 on wage regulation do not even mention it among the tools of wage control.[5] The charges for social insurance, which are borne by enterprises and which have been increased in some countries, also serve this purpose partially.[6] The taxes on wages, however, are still too small to induce enterprises to change their employment policy.

Before proceeding, the question must first be answered: why is regulation by taxation used only in a decentralised system? Taxation as an instrument of regulation can be effective only in an economy where costs have a decisive effect on the financial conditions of enterprises. This requires that enterprises be allowed to keep a portion of profit according to a formula fixed in advance, preferably through a tax on profit. Of the three countries with a classical administrative system, only in Czechoslovakia is there a surrender rate for produced profit, fixed in advance. However, no carry-over of profit is allowed; profit which is not used during the year together with profit produced above the plan (after allocations to the bonus fund) is to be surrendered to supervising agencies.[7] In the USSR enterprises are obliged to surrender what is called the free remainder, i.e. what remains after they have replenished the bonus fund, development fund and other funds, paid charges on capital and made rental payments. In the GDR the central authorities allot a certain amount of profit to enterprises.[8] Under such conditions the use of taxation as a regulator of wages would not make much sense.

In fixing taxes for the purpose of wage regulation, the centre has to decide what to tax and how to tax. As for the first point, 'what to tax', the number of options is limited to average wages or the wage-bill.

What is taxed should not be confused with the source of financing taxes. In determining what is taxed it is of relevance whether increases in average wages or increases in the wage-bill entail a tax.

It is, of course, possible to imagine (and practice confirms it) that taxes to be paid for wage increases can be financed from sources other than the wage-bill (for example, from gross income, profit or the bonus fund). All these methods have in common that they aim at braking the pace of wage increases, though in different ways. If the tax is paid from the wage-bill, the government withdraws a portion of funds which could be used for wages. In financing the tax from other sources, enterprises are really given an alternative for the use of their acquired funds. In the case of gross income and profit, enterprises are in a sense confronted with the decision whether to opt for short- or long-term interests. Namely, the money which is used for wage increases and taxes could be used for investment with the chance that this will bring about greater incomes in the future. If the tax is paid from the bonus fund (as in Hungary), enterprises have to choose between wage increases and bonuses (or other benefits available from the bonus fund).

The question 'how to tax' is really at the heart of the problem discussed in this subchapter. The question can be formulated as follows: should the tax rate be uniform[9] throughout the whole economy or at least in most economic sectors, or differentiated in order to take account of different conditions in individual economic units? If it is uniform, should the tax be proportional or progressive?

First, it must be made clear that even a uniform system may have some elements of differentation. The tax system can be designed in different ways. A tax can be levied for every percentage of wage increase over the previous year (whether it is a proportional or progressive tax) or only on wage increases above a certain limit determined by the performance of enterprises.[10] The latter possibility is applied in the present Hungarian system of wage regulation, which means that enterprises with a better performance pay lower taxes for an increase in wages by 1 per cent. Thus the tax rate for wage increases is differentiated not only by the level of wage increases but by other factors too.

As already mentioned, the underlying philosophy of a decentralised system favours uniformity. There is valid reason for assuming that taxation is the sphere where uniformity is most warranted. Taxation is an important factor of the environment in which individual enterprises work, and therefore it could be rightly argued that it should be applied equitably to all units or at least with the fewest possible exemptions.

The degree to which uniformity is applied in practice depends on several factors, probably the most important being the primary purpose of taxation: whether it has only to fulfil the regulative function of the SWR or whether it must also fulfil other functions. If the purpose of taxation is limited to one goal—to protect the economy against inflation, as it was originally intended in Hungary[11]—then strict uniformity becomes less important.[12] It is possible to imagine that regulation of wages by taxes, apart from keeping wages within acceptable limits, is also regarded as part of a whole package of taxation which is viewed as stimulus and a criterion of efficiency. If this is the case, as it was in Czechoslovakia during 1967-9, then there may be a tendency to impose a uniform tax not only on the main evaluation indicator but also on growth of wages. A differentiated tax rate (or even a uniform progressive tax levied according to the level of performance) seems to contravene the basic idea of having a tax as a test of and a stimulus to efficiency. And this was also the reason why the Czechoslovak planners, starting with January 1967, applied a uniform tax on gross income in industry and construction as well as on wage growth. It can be assumed that there will be a greater tendency to uniformity if the tax is paid from the wage-bill than from another source, mainly, if this source is profit. The size of profit does not directly depend on the number of employees, which is one of the elements decisive for the size of the wage-bill.

With this we have arrived at another question: should uniformity be reduced to a single tax rate throughout the economy or should it be combined with progressivity? One could argue as the Czechoslovak reformers did, that a proportional tax is more suitable as a stimulus to and a criterion of efficiency.[13] However, no planned economy can disregard the possible danger of inflationary pressures and the excessive widening of wage differentials. A progressive tax can be used as an important tool for mitigating both. To renounce progressivity for the sake of the pursuit of efficiency—as the Czechoslovak reformers did in 1967-8—may have the opposite effect and adversely influence price stability and wage differentials. Progressivity is all the more needed if the tax is payable from produced profit. As is known, the amount of produced profit varies considerably by enterprise, and these differences are to a great degree beyond the control of enterprises.

In this connection the question is warranted: to what extent has taxation proved itself as a regulator of wage growth, or to what extent is this an effective tool for shielding the economy from inflation? The Czechoslovak experience differs from the Hungarian; whereas in the former taxation did not prove itself, in the latter it turned out to be an effective tool.[14] The difference in results, of course, is connected with the concrete application of taxation in the two countries. In the Czechoslovak reform the tax on wage increases was supposed to be the only indirect regulator of wage growth, as it really was for some time. For reasons which have already been indicated but which will be discussed in detail in Chapter 10, the Czechoslovak planners opted for a single tax rate. As soon as the revenues of enterprises began to grow—due to the wholesale price reform—faster than envisaged in the plan, taxation could no longer fulfil its function and was supplemented by direct methods.[15] The Hungarians were from the beginning more cautious and pragmatic. Taxation has been built into the system as a second defence line, the first being the linkage of wages to an indicator of performance. In addition since 1971 the tax on wage increases above a certain limit has been heavily progressive.[16]

Up to now we have discussed taxes at length but have neglected to mention in this chapter contributions (charges) to reserve funds, particularly to the branch ministry's, as an instrument of wage control. Poland is the only country of those examined in this study which uses reserve funds in this capacity (see further, Chapter 8). The experience with this method is still limited. Nevertheless one could theoretically argue that contributions could in principle fulfil the same functions as taxes. They may even have the advantage of not acting as a disincentive against greater effort to the degree taxes do. Enterprises view taxes as an unrecoverable payment, whereas contributions to the reserve funds of the ministries are supposed to be used for wages in the future, among other things.

However, this very advantage may also be a disadvantage. Ministries may come under pressure to use the funds for wages at a time when it is not advisable to do so from an equilibrium viewpoint. In addition, if reserve funds are to be genuine reserve funds, their use is limited to the purposes and needs of associations and their enterprises under the jurisdiction of the ministry in question. On the other hand, tax yields may be used for whatever purpose the centre considers fit.

UNIFORM OR DIFFERENTIATED LONG-TERM NORMATIVES?

Administrative systems face the problem of how to encourage enterprises to abandon their old strategy of concealing reserves and of refusing to accept more stringent plans, and to extend their time horizon in decision-making. Under conditions of annual plans with binding output targets from the centre, the acceptance of demanding plans which leads to the revelation of reserves are contrary to the interest of enterprises. They fear that the central planners will take the accepted demanding plans as a basis for assigning higher output targets for the next year, which might be difficult to fulfil, not to say over-fulfil. These fears are substantiated by the behaviour of the centre. On the other hand, the logic of the system imposed by the centre obliges it (the centre) to behave in such a way. Under conditions where the central planners do not have at their disposal objective data about the productive potential of enterprises and when they know that enterprises are interested in underestimating their potential, it is understandable that the plan of the current year is the basis for the next year. The administrative system with annual binding targets make managers primarily concerned with immediate tasks and interests at the expense of the long-term.

For many years economists have been looking for ways to break this deadlock. From what has been said above, it is clear that this can be accomplished only if enterprises can be given some certainty and predictability regarding the application of economic policy by the centre, including wage policy for a considerable length of time. And this is the purpose of the long-term normatives which some countries are trying to apply. In wage and bonus regulation long-term normatives mean that during their validity the wage-bill and/or the bonus fund grow in a certain relation to the fulfilment of the success indicator, or, more precisely, that 1 per cent improvement in performance or 1 per cent over-fulfilment of certain targets entails a growth of the wage-bill (the bonus fund) by a certain fraction of the wage-bill (the bonus fund).[17] It is expected that if managers and other personnel of enterprises have the certainty that their wages and bonuses will grow according to normatives fixed in advance for several years, they will strive for a better utilisation of production factors. In order to achieve this they will have to reveal their reserves and accept more stringent plans.

In fixing long-term normatives the centre must solve two problems:

(a) Should the normatives be uniform or differentiated?

(b) It is clear that the normatives must be stable over a long period; otherwise it is pointless to talk of long-term normatives. However, the stability of normatives can take two forms: equal normatives throughout the whole period of the medium-term plan (say five years), or annual normatives, the magnitude of which is, however, fixed in advance for the whole period.

Uniformity Versus Differentiation

Before the problems raised are tackled, it should first be mentioned that long-term normatives are not applied in all the countries. Czechoslovakia nowadays does not apply them at all, in spite of being the country which first applied them (in 1959-62). East Germany applied long-term normatives for only a short period in 1969-70 and now no longer uses them. Among the countries with an administrative system, only the Soviet Union tries to apply long-term normatives for the formation of the bonus fund. (In addition some enterprises are experimenting with long-term normatives as an instrument of wage regulation.) On the other hand, long-term normatives are supposed to be a component of wage regulation in Poland.

Hungary also applies long-term normatives even though there is no mention of them in the Hungarian literature. What else, for example, can the provision be termed which allowed enterprises from 1971-5 to increase average wages by 0.3 per cent for every per cent increase in gross income per employee? In Hungary long-term normatives have a different origin and function than in an administrative system. In the Hungarian system, binding targets are assigned only exceptionally, and wage and bonus growth are not linked to the fulfilment of plan targets. Therefore it could be theoretically argued that enterprises have no interest in concealing reserves and/or in resisting an increase in effort.[18] This is all the more true because it can be expected that the market mechanism which has been granted a certain role will induce enterprises to perform more efficiently in pursuit of their own interest in maximising profit.

Long-term normatives are a logical outgrowth of a decentralised system, which can work properly only if the rules of the game (including the fixing of normatives) between the centre and enterprises are stable for a considerable length of time. The Hungarian reformers were very well aware of this principle and incorporated it in their economic reform. Up to now, most major changes in the rules of the game have occurred in 1971 and 1976, always at the beginning of a new five-year plan.

All the countries applying or trying to apply long-term normatives (the same is also true of annual normatives) rely on differentiated ones. Hungary is to a certain degree an exception. Differentiated normatives are closer to the spirit of the administrative system, because they enable authorities to have a more thorough control over wage development. In contrast to uniform normatives, they may be used as an instrument to mitigate or prevent excessive wage differentials. A system with differentiated normatives is also preferred by many enterprises, mainly the poorly performing ones. Such a system makes it possible for enterprises to press for and to obtain preferential treatment.

In the reform of 1965 the Soviet planners set as a goal the introduction of group normatives (for groups of enterprises with the same conditions) and, where possible, branch normatives. The regulations for the incentive system of 1978–80 again stress this goal.[19] It appears that the main reason for the desire to introduce group normatives is the consideration that this would exert pressure on lagging enterprises to improve their performance. Also it gives authorities a yardstick for the evaluation of enterprises. It seems that in Soviet practice normatives differentiated by enterprises prevail, something which is contrary to the intentions of central planners and regulations. As Polish experience shows, it is difficult to work out a methodology for fixing normatives for individual enterprises which would take account of all important criteria.[20] The working out of group normatives is even more difficult, as Soviet experience shows.[21] It presupposes an elaboration of reasonable criteria and the classification of enterprises into groups according to these criteria, which is a time-consuming and not a very objective job. In addition since enterprises undergo uneven changes, the grouping must also be changed from time to time.

In Hungary up to 1976 the normative was uniform for sectors of the economy and branches of industry where wage growth was linked to gross income per employee, and this was the vast majority of enterprises. The changes in 1976 which aimed at bringing the system of wage regulation closer to the conditions of individual sectors of the economy also affected the system of normatives. Nowadays normatives are differentiated by sectors and industrial branches.[22] It seems that Hungary has avoided the atomisation of normatives by individual enterprises which is widespread in the USSR and Poland. The Hungarians could afford to apply greater uniformity than other countries, since progressive taxation can take care of unwanted wage differentiation to a great degree.

Let us now turn to the second question. Should there be a single normative for the whole period of validity of the long-term normative, or should the normative change in the course of time? Of course the answer depends to a great degree on the character of the indicator and the way wages are linked to it. Suppose the indicator of performance is labour productivity or net income and the normative fixes the allowable rate of increase in wages or bonuses with an increase of 1 per cent in productivity or net income. Provided that productivity or net income is objectively calculated, there is no reason why there should not be a single normative. On the other hand, suppose the indicator is gross output and the wage-bill is the product of a wage normative for one ruble of output times value of output. In such a case there may be a good reason for changing the normative annually, since with the increase in productivity the labour costs in ruble of output decline. The Soviets who are experimenting with this method of wage regulation assign annually diminishing normatives to enterprises.[23]

Finally, to what extent are normatives really long-term? To judge from the Soviet literature on the subject, it is safe to say that the stability of normatives is more an exception than a rule. Normatives are often revised, one reason being, as Egiazarian contends,[24] that the five-year plans of enterprises have not yet become a guideline for planning production. It is self-evident that realistic plans are an important precondition for fixing reasonable long-term normatives. Central planners are able to set correct normatives only if they know what enterprises will produce and what kind of technological changes output will undergo. If the financial and output plans change frequently, as is the case,[25] normatives must also be changed.

The system of long-term normatives is thus based on the idea that it is possible to forecast correctly the development of all economic factors relevant to the indicators in enterprises. Naturally, this is an illusion. Usually many changes soon occur which alter the assumption on which the normatives were fixed. When this happens, the long-term normatives are changed, thus becoming short-term. Obviously this does not encourage enterprises to reveal reserves and adopt more stringent plans.[26]

The Polish reformers promised to link wage growth to long-term normatives, first for the period 1973–5 and later for an even longer period, possibly for the whole period of the five-year plan. Recently the Polish government has confined the new normatives to 1977–8. Hungary has the best record for stable normatives. But even there, normatives for some enterprises have been changed in 1978, i.e. in the middle of the five-year plan operation.

Part Three

7 Wage Regulation in the USSR

In this chapter the Soviet SWR is discussed in a systematic way and the GDR system briefly. It has already been mentioned that the Soviet system is a classical, direct system and that it was used by other countries of the Soviet bloc until the second half of the 1960s more or less. Therefore the analysis of the Soviet system will be at the same time also an explanation of the system in other countries in the 1950s.

THE PRESENT SYSTEM

The existing Soviet system of wage regulation came into being in the thirties, but its start did not coincide with the beginning of planning. It was not until the planners were faced with strong inflationary pressures that they started to look for effective methods for keeping the growth of average wages and the wage fund within the limits of the plan. The State Bank was soon involved in this effort as well; to it was entrusted control over wage expenditures. The linkage of the wage-bill growth to an indicator which is the main component of the Soviet SWR was introduced later.[1] In 1938, at government bidding, 210 enterprises experimented with a control of wages under which wage expenditures depended on the allotted wage-bill and the percentage of fulfilment of output targets. In 1939 the method applied in the experiment was made into a general rule.[2]

In the Soviet system wages, bonuses and employment are planned not only on the macroeconomic level but also on the microeconomic.

Planning on the microeconomic level means, as in other fields, the application of administrative methods. Up to 1966 when the USSR embarked on a reform, every level of management in the material sphere was assigned at least the following four targets which comprised the labour plan: annual average number of employed, growth of productivity, average wage and the wage-bill. The number of employed was assigned in the following groupings: wage earners, engineering technical staff, clerical personnel and junior service staff. For each group average wages were assigned.

The wage-bill of an enterprise was (and still is) calculated as a product of the planned number of employed and the planned average wage. The planning of the two components was essentially a planning of the rate of change over the previous year. This can only be carried out by planning all the factors which have a decisive influence on the number of employed and the average wage of an enterprise. In the case of the first component, this means calculating the impact of the changes in output targets on the required labour input, including changes in skill mix, in addition to giving due consideration to expected changes in productivity. In the planning of the average wage, changes arising in skill mix were taken into account, and an allowance for the growth rate of wages over the previous year was made. This was different for different enterprises in relation to the expected gains in productivity. Naturally, planning the number of employed and average wages for individual enterprises cannot be done in isolation from the rest of the economy. Planned changes in output targets have to be considered in the context of changes in the whole economy, and these, in turn, have to be checked for consistency with global manpower balance, while increases in wages must be considered in the context of the balance between incomes and expenditures for the whole population.

The reform of 1965 reduced the number of indicators of the labour plan assigned to enterprises from above to one—the wage-bill.[3] This, however, does not mean that the planners intended to give up planning the other three indicators mentioned; they are still planned on the macroeconomic level and also must be planned by enterprises. Since enterprises submit their technical-output-financial plan for approval to supervisory agencies, enterprises' decision-making even in this field has been restricted. In addition in 1971 the authorities again started to assign enterprises a productivity indicator, though in connection with the changes in the incentive system.[4] And even before this, a new indirect indicator for the employment of non-manual workers was introduced.

The wage-bill is handed down to enterprises through the hierarchical channels of management. Aside from the various experiments in different periods which will be discussed later, wage-bill growth was and is linked to gross value of output or commodity production, mainly to the former.[5] If the target in gross value is fulfilled, enterprises get the planned wage-bill. An over-fulfilment of the gross value of output target was in the past rewarded by a proportional or even higher increase in the wage-bill.[6] In 1959 the adjustment coefficient was reduced below unity[7] and differentiated by branches of industry according to the relative share of wage costs in production costs.[8] V. Maier maintains that ministries were given the right to differentiate the adjustment coefficient for over-fulfilment by enterprises.[9] Iu. Margulis, 13 years later, asserts that the adjustment coefficients are not differentiated by enterprises and, therefore, do not reflect conditions in individual enterprises—that is, the degree of labour intensity of output, the technological level, the structure of wage forms (the share of pieceworkers in the work-force), etc. What is also of importance is that the adjustment coefficients set in 1959 have not been changed since then, though the situation in individual branches of industry has undergone different changes with respect to the share of piece workers in the work force, technological level, etc. As a result the allocation of additional funds to the wage-bill for over-fulfilment is very uneven when compared with the real needs of individual enterprises.[10]

With some simplification it is possible to argue that the regulative function of the traditional Soviet SWR (i.e. the SWR in its capacity as an anti-inflationary tool) is designed in a way that generates economic inefficiency. On the other hand, the stimulative function has a tendency to engender market disequilibria. The word simplification is warranted because the two functions coalesce to a certain degree (what is regulative function is to a certain degree stimulative and vice versa) particularly in the direct system. With this qualification in mind we may start out with the regulative function. It has already been explained that the linkage of wage growth to plan targets has adverse consequences: it makes enterprises indifferent to the state of the market, encourages them to conceal reserves and makes them reluctant to accept taut plans. Each of these three consequences is, of course, a source of inefficiency. The situation is aggravated by the fact that the binding target is expressed in gross value of output. A great deal has been written on the shortcomings of this indicator. Therefore, only a brief summary will be given here.

Gross value of output by definition takes into account not only the performance (value added) of the enterprise in question but also the contribution of other enterprises, as long as it can be included in the production costs. Therefore, it is no wonder that enterprises strive to maximise the portion of gross output originating from other enterprises. There are several ways to achieve this. One is to increase the share of material intensive products in the output mix.[11] (And this is also, for example, the reason why enterprises are reluctant to produce a sufficient amount of spare parts which are usually less material intensive.) Gross value of output as an indicator creates a strange situation; the more material and energy a product contains, the easier it is to fulfil the plan. Moreover, it also encourages the use of more expensive materials and the production of more expensive products. It is also more advantageous for enterprises to buy semi-finished products and services regardless of cost rather than to produce them themselves. Cooperation with other enterprises is not considered in terms of efficiency but rather from the viewpoint of how it can ease the fulfilment of plan targets. Naturally, this indicator encourages excessive cooperation, very often beyond economic rationality.

Allocation of additional funds to the wage-bill for over-fulfilment of plan targets (stimulative function) is, of course, a great incentive to enterprises to over-fulfil targets and also an additional encouragement to them to conceal reserves. This was particularly true up to the time when the adjustment coefficient was reduced below unity.

Disregarding changes in employment, the planned increase rate in the wage-bill from one year to another was and is usually smaller than the planned increase in the gross value of output. The planners know very well the dynamics of the relationship between labour costs and gross value of output. In determining the increment in the wage-bill, they take into account potential increases in labour productivity, the percentage share of piece workers in the work force, and also consider enterprises' tendency to juggle gross value of output. However, for the sake of encouraging fast expansion of output they were willing up the 1960s—as the widespread progressive piece-rates[12] indicate—to allow an adjustment coefficient much higher than unity in some branches for a 1 percent over-fulfilment. This very willingness of the centre—which was motivated by a desire to speed up the expansion of output—encouraged enterprises to conceal reserves. Why should enterprises reveal their reserves if they could get a greater allocation to the wage-bill for the same performance classified as over-fulfilment?

From various indications and from the mentioned assertion of Iu. Margulis it is possible to conclude that even the reduction of the adjustment coefficient below unity makes over-fulfilment mostly an attractive goal to strive for. Disregarding additional bonuses and honours which such an over-fulfilment usually produces, it also confers other benefits. In some enterprises it serves as a pretext for increasing the work force and/or extending overtime work or even slackening work norms. In other enterprises over-fulfilment may result from such factors as better organisation of production and/or application of new techniques which require no increase in the wage-bill (or a smaller one than is allotted on the basis of the adjustment coefficient). Furthermore, it should not be forgotten that over-fulfilment or a part of it may be a result of juggling gross value of output (e.g., by excessive cooperation).

On the basis of the foregoing statements it seems safe to argue that the system of granting additional allocations of funds to the wage-bill for over-fulfilment bears the seeds of inflation.[13] This may even be true when the adjustment coefficient is smaller than unity, as is almost always the case when part of the remueration resulting from over-fulfilment is not matched by adequate output of consumer goods.[14] And this is not an infrequent case since enterprises in their drive for over-fulfilment of targets are not much concerned with the equilibrium effects of their activity, that is, whether the additional output would be saleable or only an addition to unsaleable inventories.

THE INCENTIVE SYSTEM OF 1965

In 1965 the USSR embarked on an economic reform which also included a reform of the incentive system. The reform did not touch the SWR. But it is fair to assume that in designing the reform, primarily the incentive system, the planners also had in mind the state of the regulation of wages. The new incentive system, which was partly discussed in Chapter 4, was marked by several innovations.

It was a partial deviation from the old system in that the dependence of the size of the bonus fund (so-called fund for material stimulation) on the fulfilment of binding targets was scaled down. The fund was no longer fixed in advance by the authorities; its size was to result from the rate of increase in sales (or profit) over the previous year and the achieved level in profitability and the normative.[15] This 'automaticity' in the fund formation aimed at boosting enterprises' initiative in accepting taut plans, which of course had to be approved by the authorities. The planners hoped to find in this new arrangement a cure for enterprises' reluctance to accept demanding plans and thus to reveal reserves. Why should enterprises be reluctant to change their behaviour and not accept taut plans, if they have the assurance that for every increase in sales (or profit) and for every percent of achieved profitability they will be given an allocation to the bonus fund? Why should the centre impose higher and higher targets every year if enterprises are willing to improve their performance out of sheer material interest?

In order to be surer that the new incentive system would act in this direction, the planners applied two further provisions, both new. The new system penalised both over-fulfilment and under-fulfilment of the plan. If an enterprise over-fulfiled both or one of the so-called fund-forming indicators, the normative for that portion which exceeded the plan was reduced by at least 30 per cent. A roughly similar disincentive was specified for under-fulfilment of plans.[16] This provision aimed not only at encouraging enterprises to accept demanding plans but also at discouraging them from committing themselves to unrealistic plans. The second provision involved the introduction of long-term normatives which meant a pledge on the part of the planners that the normatives would not change annually. It was hoped that such an assurance would help to extend the managers' time horizon in decision-making.

The type of indicators chosen gives evidence not only of a new approach to incentives but also that a certain amount of thought was given to the idea of offsetting the adverse effects of linking wage growth to output targets. In a sense they represented a combination of quantitative (of the two alternatives of the first indicator—sales or profit—most enterprises used sales)[17] and synthetic indicators. What was also of importance was that profitability was assigned an important role and was defined in a new way. Usually profitability was defined as ratio of profit to production costs; for the first time it was calculated as a ratio of net profit to invested capital. Profitability thus defined had to induce enterprises to become more efficient by a better utilisation of capital among other things, thereby countering the effects of the quantitative indicators.[18]

The new incentive system was short-lived; in 1971 the authorities carried out changes in the system which meant in substance a return to the old methods. The fund-creating indicators which were—as already mentioned— supposed to ensure some automaticity in the formation of the bonus fund were downgraded to incentives for over-fulfilment of targets. As was noted in Chapter 4, the bonus fund is again assigned by the centre as an absolute sum. The new incentive system was brought down by several factors. Some of these were in the system itself, and some had to do with the environment in which the new system had to work. Last but not least, the fact that the system did not live up to expectations strengthened the hands of those *apparatchiki* and economists who, from the beginning, opposed 'playing' with reforms.

It would exceed the scope of this study to dwell extensively on the reasons for the failure of the incentive system. Therefore only some of the reasons will be briefly sketched. Productivity targets, set apparently in the optimistic expectation of an upturn in the economy, were not wholly fulfilled.[19] What was also annoying was that average wages in industry grew faster than envisaged in the plan. Many authors blamed the design of the new system for the results in productivity.[20] The sales indicator was undoubtedly not a strong stimulus for increases in productivity. Yet the linkage of the size of the incentive fund to the size of the wage-bill—a provision motivated by the desire to avoid great differences in the bonus fund—acted as a disincentive for productivity growth. Since managers received a larger share from the bonus fund than workers, the new incentive system provided them with an incentive to strive for a higher wage-bill and employment at the expense of productivity.

Profit and profitability can be relatively objective indicators provided the price system is rational and there are no great differences in the profitability of groups of products. True, the wholesale price reform of 1966–7 was designed so as to reduce large and unjustified differences in profitability.[21] However, it fulfilled this task only partially.[22] Some branches left with small profitability by the reform (e.g., coal, ferrous metallurgy, forestry) got into a disadvantageous position with regard to the bonus fund.[23] They undoubtedly pushed for changes.

Normatives which were supposed to play an important role in the incentive system, as already indicated in Chapter 6, were reduced to not much more than a formality. The planners encountered tremendous difficulties in setting what were henceforth to be group normatives, all the more because enterprises were not all converted to the system at the same time. The normatives were often changed, in many cases in a way to ensure enterprises a certain bonus fund.[24] According to G. Egiazarian the system of setting normatives introduced elements of egalitarianism into the formation of bonus funds in some branches and this undermined the effectiveness of the normatives.[25]

As has already been shown the incentive system adopted in 1965 went through several modifications, major ones in 1971 and some smaller ones in 1976. The modified system has retained some devices for stimulating enterprises to accept taut plans, to increase productivity and to care more for consumer demand and quality of goods. The incentive fund is planned for the whole period of the five-year plan. For overfulfilment of annual targets (set in the five-year plan) for productivity growth, for output of goods of popular demand and for high quality goods, additional funds are allotted to the incentive funds.[26]

EXPERIMENTS IN WAGE REGULATION

Disappointment with the results of the reform of 1965 increased interest in the regulation of wages. This is reflected in the intensification of experiments which proceed in two directions. The Soviets are experimenting both with net output and with the linkage of the wage-bill to a wage normative for one rouble of output. The purpose of Soviet experiments with net output is not only to find a new indicator for wage growth regulation but also to find a new yardstick for evaluating enterprises' performance and a new, more objective numerator for the computation of productivity. This latter purpose alone has given a boost to the ongoing debate on this topic and has recently spurred *Voprosy ekonomiki* to open its columns to a debate which was introduced by E. Kapustin.[27]

The first stage of the experiment with net output was carried out in 1969–70. Though experimenting enterprises and planners found the new indicator useful, they also encountered many accounting difficulties. It seems that the main difficulty stemmed from the lack of figures for material costs in constant prices. Net output was defined as the gross value of output reduced by material costs. Without having figures for material costs in constant prices, they had difficulty comparing the dynamics of the gross value of output with net output.[28]

In 1973 the experiment was extended, and in order to avoid the accounting difficulties mentioned, experimenting enterprises started to use the so-called 'normative net output'. It is called normative since the value added of individual products which make up the net output of an enterprise are expressed in terms of stable normatives (possibly for five years) as part of the wholesale prices of products. For example, let us suppose that the wholesale price of a product is set at 20 roubles; its normative net output is, say, 75 percent; in money terms this means 15 roubles. What is of relevance is that this net output price is stable whatever the changes in material costs may be. This means that changes in the material costs of individual products are neutralised; what really counts are changes in the volume and structure of output.[29] This is not to say that the adherents of net output claim to have found a panacea for all the shortcomings of gross value of output. They are aware that net output will create new loopholes and cannot be used for all purposes. On the whole, however, they find it a more objective indicator than gross value of output.[30]

Some economists advocate their stand in favour of net output by arguing that enterprises currently experimenting also over-fulfil targets in spare parts and are more willing to produce goods of popular demand.[31] Such a development could be expected since the new indicator encourages labour intensive products rather than material intensive. In particular net output encourages a shift to more profitable products, which is made possible by the distorted price system. Since profit has greater weight in net output than in gross output, shifts in output mix to more profitable products are felt more in net output than in gross output. Precisely this greater sensitivity of net output is used by its opponents to argue that productivity computed with the help of net output would show up greater increments than if it were computed with gross output. For the same reason they argue that a linkage of the wage-bill to net output would push up wages without a proper increase in output and consequently would be inflationary; therefore they object to it.[32] Some respond to these objections against net output by arguing that the shortcommings result from prices and not from the indicator itself.[33,34]

Let us now turn to the second, relatively widespread experiment—the linkage of the size of the wage-bill to a wage normative for one rouble of centrally planned output, which, it appears, has already outgrown the experimental stage in some branches. This new method of wage regulation (used experimentally only by manual workers in industrial activity) is not applied in all branches in exactly the same way. The common characteristics can be summarized as follows.

The wage-bill is the product of the value of output of the enterprise times the wage normative for one rouble of production.[35] In experimenting branches, the yardstick used for calculating production is usually gross value of output or commodity production. In most cases the normatives are supposed to be long-term, diminishing annually in the planning period. It is assumed that wage costs per unit of production will decline due to increasing productivity. This is, for example, the way the experiment was designed in the ministry of instrument making and automation during the quiquennium 1971-5.[36] In the chemical and petroleum machine building ministry the normative method applied is more complicated. The normative also diminishes annually, but its rate of decline depends on the way the expansion of production is achieved. The more it is achieved through productivity growth, the higher the so-called normative coefficient which determines the wage normative for one rouble of production.[37, 38]

How is plan over-fulfilment treated in this experimental method? Two kinds should be distinguished: first, the accomplishment of a greater reduction in labour inputs and thus in wage costs per unit of production than envisaged in the plan and, secondly, the over-fulfilment of output targets. Some of the savings due to higher productivity can be, as in the Shchekino experiment, used for increases in wage rates and for one-time rewards to workers.[39] It is, however, not clear how much can be used in this way. Over-fulfilment of output targets is rewarded, but in the timber and wood processing ministry, for example, the wage normative for over-fulfilment is lower than for fulfilment of plan targets.[40]

In comparison with the traditional method, the normative method of wage-bill regulation means that wage growth is more closely connected to performance. Though it also means a greater automaticity in the formation of the wage-bill it remains in the direct SWR framework. It is hoped that the new method will stimulate enterprises to adopt more demanding plans and pay greater attention to the economic use of labour and thus promote productivity. Undoubtedly the success of the new method will largely depend on the extent to which the savings gained are allowed to be used for wage purposes. There are two constraints to a generous policy in this regard, as experience with the Shchekino experiment has shown. Planners are afraid that permission to use a large portion of savings may, as already indicated, bring about inflation and in some regions cause employment problems.[41]

BANK CONTROL

It has already been mentioned that the State bank (*Gosbank*) monitors the disbursements of wages, an activity which has a long history. As early as 1931 all enterprises were obliged to inform the Bank about the amount of the authorised wage-bill for individual quarters of the year, broken down by months. The actual payments of wages were not supposed to exceed the planned. It was the Bank's duty to report over-expenditures of the wage-bill to supervising agencies whose obligation it was to initiate corrective measures or even in extreme cases to prosecute the people responsible for the overdraft. In 1933 the rules were tightened; over-expenditure of the planned wage-bill was allowed only if the output plan was over-fulfilled.[42] However, there was not yet in operation the well-known rule linking the wage-bill size to the fulfilment of output targets, a development occurring in 1939, as already mentioned.

The new rules of 1939 for drawing funds were limited to industry and were not a very effective tool against excessive wage expenditures. First, output targets were calculated in gross value of output, a yardstick which can be—as has been shown—easily manipulated. Secondly, for an over-fulfilment of output targets enterprises were allowed to over-expend their wage-bill proportionally. And finally, the Bank was obliged to provide funds for over-expenditures of wages up to 10 per cent in the first month. According to some writers this provision was in practice interpreted in a way that allowed enterprises to overdraw their wage-bill every second month; that is, every month which was preceded by a month in which the enterprise in question did not overspend the wage-bill. In addition enterprises could ask supervising agencies for permission to go beyond the above-mentioned limits, and this was handled leniently.[43]

The rules of 1939 introduced relative stability into the criteria for disbursement of wages and Bank control. Aside from the ongoing experiments, the linkage of wage-bill growth to gross value of output is still in force. The provision regarding the allocation of additional funds for over-fulfilment of output targets was not changed until 1959, when the adjustment coefficient was set below unity. Reference has already been made to the fact that even such an arrangement may be inflationary. Therefore Soviet authorities have recently been trying to bring the allocation of funds to the wage-bill for over-fulfilment of output targets in line with productivity. For this purpose the Bank is empowered to reduce the adjustment coefficient up to 50 per cent if the over-fulfilment is achieved by employing more workers than planned and thus productivity targets are not fulfilled.[44] However, the provision concerning the possible overdraft of funds by 10 per cent in the first month was dropped, probably after the Second World War. V. Popov's book on the *Gosbank*, which also discusses the control of expenditures for wages and which was published in 1957, does not mention such a possibility.[45]

The reform of 1965 brought about little change in the Bank control function, except that it was made more flexible by replacing monthly controls with quarterly.[46] However, this modification has recently been partly reversed. Due to many complaints[47] enterprises with recurrent overdrafts have again to undergo monthly controls.[48] The reform also meant a change in the criteria for disbursement of wages. Up to the reform under-fulfilment of output targets entailed a proportional deduction in the wage-bill. With the reform the adjustment coefficient has become applicable for under-fulfilment too.[49]

In their effort to induce managers to observe wage plan targets, the Soviet authorities also use penalties. It is not clear when they were applied for the first time. At any rate the regulations for granting bonuses to employees of enterprises which converted to the new incentive system of 1965 already contain such a provision. According to this, the bonuses of top managers can be reduced up to 50 per cent if the wage-bill is overdrawn. However, if they manage to redeem the overdraft within six months they are entitled to a refund of 50 per cent of the penalty.[50]

To our knowledge there is not enough information to make possible a clear-cut judgement on the effectiveness of Bank control. What is clear is that the existence of the control does not of itself prevent enterprises from an over-expenditure of the planned wage-bills. The main complaints in publications about overdrafts[51] as well as the recent tightening of the rules of control are the best proof of this. It is, however, not known how effective the Bank's interference is once overdrafts are discovered. In the 1960s one Soviet economist maintained that 80 per cent of the overdrafts were compensated for by a better performance within the set time limits.[52] We have not come across new figures on this topic.

WAGE REGULATION IN THE GDR

The GDR is the only country to which no separate chapter is devoted. First, there is surprisingly little literature on the specifics of wage regulation.[53] Secondly, from what is known, it seems that the GDR has in substance followed the Soviet pattern of wage regulation. In 1970 a resolution was adopted to change the system somewhat but this intention was soon dropped.

As in the USSR, the wage-bill in the GDR in the fifties was assigned to enterprises by the centre. The planned wage-bill could be disbursed if the plan target for gross value of output was fulfilled.[54] It is not clear from the legislation how over-fulfilment of the plan target was treated and how high the adjustment coefficient was. It is worthwhile mentioning that the German planners, it seems, devoted more attention than other countries to the avoidance of an over-expenditure of the wage fund. For this purpose enterprises were put under the strict control of the National Bank.[55] Of no less importance was the fact that the central authorities put great stress on the meeting of employment targets. An over-fulfilment of employment targets could be penalised by payments of DM500 from profit for each worker above the planned number, if corrective measures were not undertaken within a set time.[56]

In the late 1960s,[57] in the wake of the Seventh Party Congress resolution in 1967, the Germans experimented[58] for some time with linking the wage-bill to productivity on the basis of a normative. For this purpose the wage-bill was divided into two parts: the basic wage-bill and its increment. The basic part was assigned by the centre and corresponded to the previous year's size, modified by changes in the planned number of employed and in the qualification mix. The increment was linked to the planned increase in productivity on the basis of a normative. It is not clear how productivity was calculated.[59]

In 1970 the central authorities decided to link increments in the wage-bill to long-term normatives for the period 1971–5. This was not an isolated action; it was part of a whole package of modifications for the new five-year plan designed to extend the application of economic instruments. The growth of the bonus fund was also supposed to be linked to long-term normatives. The government proclamation regarding these changes and the decree which implemented them were published in 1970,[60] but at the end of 1970 the German authorities abruptly reversed the course of the economic reform, a reversal which was also reflected in the dropping of the idea of long-term normatives. There is no agreement among economists about the reasons for this change. Some look for them in the economic field, mainly in the structural problems the GDR faced in 1969–70 and in the decision, which followed, to reduce the objectives of the plan.[61] K. Hensel sees the main reason in the political sphere. According to him the difficulties which affected the economy enabled those in the highest echelons of power who opposed decentralisation to gain the upper hand.[62]

Judging on the basis of legislative provisions and other limited sources,[63] the situation in wage regulation is the following. The wage-bill assigned to enterprises by supervisory bodies can be fully utilised 'if productivity is increased according to the plan and the state target in commodity production is fulfilled with the planned number of workers'.[64] From an article on the working of Bank control it seems more than probable that what the planners mean by productivity is really commodity production per employee.[65] If the preconditions are not met, it is left to supervisory bodies (in some cases to the appropriate ministry) to make decisions about the size of the wage-bill. In their decision-making the supervisory bodies must adhere to one principle—the need to encourage higher productivity, a specially burning problem in the GDR due to shortages of labour. For this purpose they are entitled to allow enterprises to use for the wage-bill savings which result from fulfilling commodity production with a smaller number of workers than envisaged in the plan. The legislative provisions do not preclude the possibility of allowing all the savings to be used. It is not known what happens in practice; but it is doubtful whether the planners would allow great increases in wages due to savings.

What is of special interest is that not every over-fulfilment of the plan target in commodity production is rewarded by an additional allocation to the wage-bill. Only over-fulfilment of targets in goods which are of importance to society gives enterprises a claim to an overdraft of the planned wage-bill. In such cases exceeding the planned working time is also allowed. No adjustment coefficient is set; apparently it is left to the supervisory bodies to determine the size of the additional allocation to the wage-bill.

An overdraft of the wage-bill which is the result of unsatisfactory work by an enterprise is penalised. According to the provisions the year-end bonus of top managers can be reduced by 20–50 per cent depending on the extent of the overdraft.[66] The control of the disbursement of the wage-bill and the decision whether an enterprise has overdrawn its wage-bill are in the hands of the Bank.[67]

8 Regulation of Wages in Poland

Among the countries examined, Poland has been most frequently plagued by inflation. It therefore deserves special attention. It is also the only country which did not take part in the wave of reforms in the second half of the 1960s. However, it did make all the preparations for a reform which was supposed to go into effect in January 1971 but which was abandoned due to the unrest which broke out in December 1970. In 1973 Poland finally embarked on a reform which included a new SWR. This chapter will concentrate on the present SWR which came into being as a result of the reform. However, for the reasons mentioned above we will first discuss the Polish system in the 1960s. This will also shed some light on problems which could not be dealt with in other chapters.

REFORM IN THE 1960s

At the beginning of the 1960s Poland followed the Soviet system of wage control quite closely. Wage growth and employment were directly regulated by the centre. The wage-bill was assigned to enterprises along with output targets. If enterprises fulfilled the plan for gross value of output or commodity production, depending on the type of wage-bill regulator used, they were entitled to the planned wage-bill. Its planned magnitude could be increased if enterprises voluntarily accepted higher targets and fulfilled them. They also had to fulfil the target for average wage growth in relation to productivity growth.[1]

For an over-fulfilment of the output indicator (gross value of output or commodity production) an enterprise was entitled to an additional allocation to the wage-bill. The adjustment coefficient was differentiated; in industry it amounted on the average to 0.5.[2] An under-fulfilment of the output indicator entailed a reduction in the wage-bill.

Enterprises are interested in maximising the wage-bill.[3] In order to discourage them from such behaviour, the system had two built-in incentives—one positive and indirect, the second negative and direct.[4] The fact that the size of the factory fund[5]—from which year-end rewards were paid—was linked to the financial performance of the enterprise was intended to be the positive incentive. It was hoped that this linkage would induce managers to strive for a reduction in wage costs per unit of production as a means of improving the financial performance of enterprises. The negative incentive lay in the sanctions which could be applied in the case of an over-expenditure of the wage-bill. If the State Bank found the over-expenditure unjustifiable, up to 50 per cent of the bonus fund of white-collar workers could be suspended.[6]

The incentive[7] and disincentive[8] did not prove strong enough to keep wage growth within the limits of the plan. Wage-plan over-expenditure was not a new occurrence; Poland was plagued by it almost permanently. At the beginning of the 1960s, when economic growth slowed down, when employment grew faster than envisaged by the plan[9] and when over-expenditure of the wage fund was not matched by corresponding increases in the output of consumer goods, the adverse consequences of the existing system were felt more strongly. It should be borne in mind that the tendency to maximise the wage-bill encouraged over-fulfilment of output targets regardless of whether the additional output could be sold or would only mean an enlargement of unsaleable inventories. Also in many cases the over-fulfilment of the plan was achieved by juggling output mix without a real need for additional wage funds.

In order to cope with mentioned problems at least partially, in December 1963 the Polish government changed the provisions for allocations to the wage-bill for over-fulfilment of plan targets. Whereas the system in existence up to that time could be characterised as an automatic system, the new system which was applied to only approximately 50 per cent of all industrial enterprises[10] was to be based on economic analysis. Over-fulfilment of output plans was to be limited to cases where economic criteria warranted it. For permitted over-fulfilments, allocation of additional funds for wage-bills were financed from the special reserve funds of associations and ministries. These funds could also be used for higher wage expenditures arising from an increase in output of products of higher labour intensity than the plan envisaged, as long as the output could be used for export or for the domestic market. The ministries set the guidelines under which the over-fulfilment of output plans would be justified, whereas associations made the concrete decisions in response to proposals by enterprises. Yet associations could also initiate over-fulfilment of output plans if they deemed it economically justifiable. A portion of the reserve funds was kept in the hands of ministries, and the balance was distributed among associations. First the association's reserve fund was used; only when this was exhausted could that of the ministry be used. This arrangement had to serve as an alarm system. Whenever an association overspent its reserve fund for goods earmarked for the domestic market,[11] the signal system went into action. The Bank informed the ministry and pressed for corrective actions.[12] The reserve system was introduced for the purpose of: avoiding unjustifiable over-expenditure of wage-bills, limiting over-fulfilment of output plans to output which was saleable and avoiding unjustified increases in labour intensive output which would contribute to employment growth.[13]

To what degree was the new system successful? No doubt it decreased the amount of over-expenditure of the wage fund, but only temporarily.[14] The 'reform' did not touch the main defects of the SWR, namely, the linkage of wage growth to a planned output target and the fact that the success indicator was gross value of output. Though initially the reaction to the 'reserve system' was positive, it soon changed as the impact of the bureaucratic offshoots of the system started to be felt. The reserve system was based on the assumption that associations had the needed information and political power to make correct and impartial decisions. As could be expected, this was not the case.[15] In addition the system enhanced bureaucratic paperwork with all its consequences.[16] The reserve system was gradually eroded; in 1969 only 22 per cent of industrial enterprises were still under the system.[17] Apparently the others managed in various ways (application of political influence not excluded) to convince the authorities to let them return to the automatic system.

THE ABORTIVE REFORM OF 1970

Before examining the 1973 reform we will briefly discuss the reform draft of 1970. This is a worthwhile endeavour since that reform intended to apply a method not previously used in other countries. The reform of the SWR was supposed to help narrow the widening gap between growing purchasing power and inadequately expanding output of consumer goods and services by restricting wage growth and holding down growth in employment, a policy already pursued earlier.[18] (It was to be supported by the price increases of December 1970 which, under pressure of popular resistance, were rescinded.) For this purpose the wage increases of wage earners were to be linked primarily to the growth of the salaries of white-collar workers, while the salaries of the latter were to be limited to increases in the variable part of their salary, i.e. bonuses.

Their bonus fund was tied to one of four synthetic indicators—reduction in costs per unit of production, reduction of the final costs of production, profitability, or amount of profit—and to many special indicators.[19] Its rate of growth was supposed to be differentiated by enterprises and was to amount in 1975 to a maximum of 80 per cent over the level of 1970. The planners wanted to use this growth rate differentiation for two purposes: to help correct disparities in differentials and to reward preferentially enterprises which distinguished themselves as pioneers in advancing technological progress.

The basis for the wage earners' wage-bill was to be the one for 1970, adjusted downwards for unjustifiable over-expenditures and upwards for allowed increases in employment. Increases in the wage-bill were to come from four sources, the main one being a quota which depended on the increase in the salaries of white-collar workers. For every 1 per cent increase in salaries, the wage-bill of wage earners was supposed to rise in the range of 0.5–0.8 per cent. In addition savings resulting from a reduction in employment in relation to the plan target, and bonuses withheld from managers for not fulfilling some tasks, could also be used for wage increases.[20]

The idea of linking wage increases to increases in the bonuses of white-collar workers was new. Of course its main purpose was to restrain wage increases. Since bonuses constituted approximately one-sixth of salaries, the 80 per cent allowable growth in bonuses already referred to meant that wages and salaries were supposed to rise by a maximum of 15 per cent in five years.[21] The abortive system also contained two other novelties which had already been applied in other countries: one, the long-term planning of bonuses and, secondly, the linkage of bonuses to performance instead of to plan targets.[22]

THE REFORM OF 1973

The unrest of December 1970 was a clear indication of the workers' dissatisfaction with the existing scale of priorities in the economic policy. Throughout the 1960s Poland had the lowest rate growth of real wages in the Soviet bloc (1.8 per cent). The new leadership under Gierek well knew that a reorientation of priorities was needed as well as a change in wage policy if the workers were to be appeased. This new attitude to wages has been reflected in the fast growth of wages in recent years. Naturally a rapid growth of wages can be ensured in the long run only if the economy grows fast and its efficiency is enhanced. To achieve this the Gierek government committed itself to carrying out a reform which was put into effect in January 1973.

The new reform was not a one-time operation; it started as an experiment in several newly created associations and their enterprises, and it has gradually been extended to other associations. Since the reform has substantially increased the role of the new associations, in the Polish literature the whole reform is termed **WOG** (*Welkie organizacje gospodarcze*—large economic organisations) system of management.[23] Many enterprises have not yet been converted to the new system which involves changes in planning, in the administration of foreign trade and in the financing of investment, as well as in the regulation of wages and bonuses.

In the new system wage-bill growth is linked to output added, more precisely to the increment in output added in industry (and to net income in trade), while in construction it is set as a proportion of output added. The following formulae are used for the computation of the disposable wage-bill:[24]

$$F_n = F_o \left(1 + R \frac{P_n - P_o}{P_o} \right)$$

$$F_n = P_n \frac{F_o}{P_o} U$$

where F_n is the disposable wage-bill;

F_o the wage-bill of the previous year adjusted for comparative purposes;

P_n and P_o output added of the current year and of the base year respectively;

R incremental normative, expressing the rate at which the wage-bill increases with the increase in output added by 1 per cent and

U proportional normative.

Output added is defined as the value of output expressed in realised prices (i.e. in the prices enterprises get for their products), reduced by the value of used or purchased materials, used energy and services, annual investment loan repayment and appropriate interest payments on investment loans and the turnover tax, and increased by subsidies. The given definition is a generalisation of the concept of output added as it was practised in 1973. Associations which later converted to the system applied a definition of output added which is different in some aspects.[25]

The supervising ministry sets the normatives 'R' and 'U' in a differentiated way to associations, and the latter have the right to differentiate them for enterprises under their jurisdiction. 'R' is usually set in the range of 0.5–0.9 whereas 'U' is in the range of 0.95–0.99. Since 'R' is smaller than unity, this means that the increment in output added is divided in a way which allows profit to grow faster than wages. This also causes the share of the wage-bill in output added to decline. In enterprises which apply the proportional formula, the normative is supposed to diminish annually; consequently the share of the wage-bill in output added also diminishes.[26] 'R' and 'U' are designed as long-term normatives. Initially they were set for three years (1973–5) with the intention of setting them for five years at the start of the new five-year plan period. However, this intention did not materialise, as will be shown later.

The disposable wage-bill is the sum of funds earmarked for wages and bonuses (not for managers) to which enterprises are entitled on the basis of their performance. The real outlay on wages may be smaller or higher than the disposable wage-bill whose size can be known only after the end of the year when the amount of output added can be figured out. If it is smaller (higher will be discussed later) the balance is to be used for reserve funds. According to a provision of 1974,[27] 35 per cent of the balance is to be transferred into the association's reserve fund and the remainder to the enterprise's reserve fund.[28] In some enterprises a portion of the remaining 65 per cent is used for 'interim' bonuses.[29]

Apart from the disposable wage-bill there are other separate funds: factory fund for rewards (from which year-end rewards are paid), managerial bonus fund and funds for stimulating technological progress, rationalisation, etc. The year-end rewards were introduced in 1956 in order to boost the personnel's interest in the overall performance of their enterprise. Originally the size of the fund was linked to the fulfilment of success indicators,[30] an arrangement which was abolished in units converted to the new system. At present its size is defined, as before, as a percentage of the wage-bill, and it is to grow on the basis of a long-term normative to a maximum of $8\frac{1}{2}$ per cent of the wage-bill, which would amount to an additional month's wage for enterprise personnel.[31] It is assumed that this will be gradually achieved in the whole industry by 1981.[32] In contrast to the reform of the wage-bill which has been applied only in units which have adopted the new system, the bonus fund for managerial staff has been applied to the whole industry since 1 January 1973.[33]

As has already been noted, the bonus fund is fed from profit and its size depends on the amount of produced net profit.[34] In order to preclude a rapid growth, the fund is subjected to a steep progressive tax, but enterprises can alleviate taxation by transferring a portion of the increment in the bonus fund to the reserve fund for bonuses. This portion is tax exempt and can be used for bonuses in years of poor performance.[35]

The old SWR was subject to criticism for many reasons. Due to a lack of objective criteria, the wage-bill assigned to enterprises was of necessity largely the result of arbitrary decisions made by the centre. The system also opened the door to bargaining and 'lobbying' by enterprises at the centre where economic arguments were not always decisive. What is more important, the system generated thriftlessness and inefficiency for several reasons. The linkage of wage-bill growth to binding output targets expressed in gross value of output was only one of these factors, though no doubt the most important one. Another was the strict control of wages from the centre and the policy of small wage increases, which proved to be a disincentive to productivity growth. Therefore managers looked for and usually found ways to increase employment.[36] The combination of a policy of low wages with a fast increase in employment closed the Polish economy in a vicious circle. Low wages hampered growth of productivity and led to expansion of employment beyond economic rationality, and a high rate of employment became an impediment to a faster growth of wages.

The intention of the reform has been to allow the Polish economy to break out of this vicious circle. It is expected that by giving enterprises greater autonomy in matters of remuneration and employment, combined with a shift to a faster growth of wages, new incentives will be created for higher productivity and a more rational employment policy on the part of enterprises.[37] The new method of wage-bill regulation differs from the old not only in the change of indicator but also, what is even more important, in the fact that the size of the wage-bill is no longer linked to the fulfilment of planned output targets. It is linked to the performance of enterprises; the size of the wage-bill in a given year is dependent in industry on the increment in output added compared to the previous year. A new important element of the new SWR is its attempt to make 'R' and 'U' long-term normatives.

The economic significance of these two measures has already been explained (see Chapters 3 and 4). Here it need only be repeated briefly that the linkage of the wage-bill to performance has to induce managers to pay greater attention to consumer preferences and to adopt more demanding but, at the same time, more realistic plans. It is hoped that giving managers a sense of greater certainty and predictability will make them more interested in long-term strategy instead of in the short-term one to which they are accustomed. All the countries under review are trying to solve the problems mentioned; however, the methods they use are different, depending on the system of management. One can argue that Poland has chosen a middle-of-the-road solution.

The provision making the size of the wage-bill dependent on performance does not mean that the wage-bill ceases to be centrally controlled. Control is retained through the assignment of the normatives 'R' and 'U' and more recently—as will be shown later—through charges to the branch reserve fund as well. The Polish reform has stopped short of the Czechoslovak (1966-9) where wage growth in most of the enterprises depended on produced gross income and was, for a certain period, regulated indirectly by taxation (instead of a normative).[38]

The new Polish reform is marked by a new approach to bonuses. In the countries of the Soviet bloc in the 1960s, the economic reforms set as a goal the strengthening of incentives, and this was to be achieved by expanding the weight of bonuses. For example the Czechoslovak reform envisaged a permanent increase in the weight of variable components (bonuses, year-end rewards) in earnings. The Soviet reform of 1965 was also intended to go in this direction, and until recently the same could be said of the Hungarian reform. It seems however that the Polish reform is moving away from this concept. The provision mentioned, that interim bonuses after two years can be integrated into wage rates, is a good indication of the direction of the reform. Apparently the reformers prefer increases in the fixed components of earnings to the variable ones.[39] Some disappointment is felt with bonuses as incentives.[40]

Of great importance for the effectiveness of the new system is the degree to which the indicator of performance can be made objective.[41] The more objective the indicator, the more effective the wage control system can be, also because it will induce managers to behave more rationally. Objectivity depends to a great degree on the definition of the indicator, on what is and what is not included in it.

The possible definition of output added has been the subject of a long debate in which three issues in particular were paramount: should the used or the purchased materials be deducted; should the turnover tax be excluded, as the original definition promulgated, or included; and finally, would it be more advantageous to deduct capital depreciation instead of instalments including interest on investment loans? Most of the associations which converted to the new system in 1973 opted for a deduction of purchased materials.[42] Nowadays almost all associations deduct the value of used-up material from output added.[43] They have come to realise that this solution benefits them more since the other one puts indirect limits on the amount of purchased materials. If purchased materials grow faster than output added, they necessarily affect unfavourably the disposable wage-bill. However, it is essential to have a sufficient stock of materials in situations where it is hardly possible to expect a smoothly functioning procurement of materials.

In order to neutralise the possible impact of the turnover tax on output mix, the planners excluded it from output added. However, experience has shown that this goal has not been fully achieved. As could be expected, a shift in output mix to products with lower tax rates has occurred. This has contributed to a faster growth of output added than of sales which contain the tax. Hence some authors argued in favour of including the turnover tax in output added. According to them a great difference between the growth of output added and sales would necessarily reduce the objectivity of the evaluation indicator and might cause inflationary pressures.[44] This view is not generally shared, and most economists maintain that the existing arrangement should be retained. They express fear that a change might induce enterprises to shift to products with a high rate of taxation and thus weaken enterprises' interest in output for export (exported goods are not subject to taxation) and in output of essential goods for the domestic market.[45] It seems that an inclusion of the turnover tax in output added would also strengthen enterprises' interest in a shift in product mix to more profitable goods. Under the existing conditions of tax collection at the level of productive enterprises, no ideal solution to the problem exists; whatever approach were chosen, enterprises would be able to take advantage of it. The only solution would be to reform the turnover tax, to make it more even. However, such a reform is supposed to be introduced only step by step in the future.[46]

Some economists argued in favour of deducting capital depreciation instead of annual instalments including interest on investment loans. They maintain that the size of depreciation is a better gauge of the size of deployed assets since it is stable in contrast to payments on loans which may fluctuate. In addition depreciation as a component of production costs induces enterprises to a better utilisation of invested capital.[47] Since in addition some associations complained that the deduction of interest and annual instalments on investment loans unfavourably affected their disposable wage-bill, the authorities resorted to the following compromise: instead of annual instalments on investment loans, depreciation is deducted. As for interest payments no change occurred.[48]

The exclusion of material costs from output added is self-explanatory if the planners wished to have output added as an indicator. Interestingly enough they also (except where the length of the production cycle warranted it)[49] excluded work in progress and changes in the inventories of finished goods, something which even the Czechoslovak reformers did not do with regard to gross income. Output added is conceived in Poland to be a sales indicator rather than a value added indicator. The rationale of this approach is obvious; the planners want to bring to a halt the hoarding of inventories and prevent enterprises 'doctoring' work in progress.

MODIFICATIONS OF 1976–7

A proper definition of the evaluation indicator is not a sufficient condition to make wage control an effective anti-inflationary tool. There are other circumstances which might turn the indicator into a non-objective yardstick of performance. One of these is the manipulation of output added by the manipulation of prices. The planners have not been very successful in preventing such activities. Wholesale prices of old products cannot be changed except with the explicit approval of the central authorities. On the other hand, new products provide a good opportunity for influencing output added. Prices of such products are approved by the supervisory body on the basis of proposals, supported by cost calculations, made by productive enterprises. Very often the process is nothing more than a formality; the supervising authority has usually no way to verify the correctness of the calculations. Therefore enterprises can get away with price increases which are not warranted by the additional costs. Many enterprises achieved higher output added simply by shifting to more profitable products while others managed to turn their involvement in foreign trade under inflationary conditions to their advantage. On the other hand, the lack of a logical link between foreign market and domestic prices brought great losses to some organisations.[50] In brief, the size of output added has lost much of its effectiveness as a yardstick of performance. This of course does not add to market stability and may be a source of disparities in wage differentials.

Employment in units which adopted the new system did not develop as was expected. The planners believed that the linkage of wage growth to output added would lead associations to slow down their employment growth, but instead employment grew there twice as fast as in units with the old system. There were some good reasons for a faster growth of employment in some units. The new associations shared to a greater degree than the old in the huge investment activities of Poland, and in addition they established institutes of research of their own. Analyses of the unfavourable development in employment also showed that associations disposed of sufficient revenues to raise the wages of their workers at an acceptable rate so that they were not under pressure to be concerned with employment. Also, the old tradition of hoarding labour for rainy days is still in existence.[51]

All this has been occurring at a time when Poland is eager to bring about a gradual reversal in its balance of payment deficit by accelerating the rate of export growth and by damping the growth rate of imports. Yet it is not an easy task to increase exports—needed to pay off the huge credits received from Western countries for the modernisation of industry—in view of the present inflationary plight of the West.[52]

One of the main objectives of the reform of 1973, to achieve higher productivity through higher wages, has been accomplished to a great degree. Plan targets in productivity were over-fulfilled in 1971–5. Yet, average wage growth exceeded the plan target by an even greater rate.[53] Poland witnessed a considerable acceleration in wage growth which probably added to the inflationary pressures.

These were some of the important circumstances which spurred the government to introduce some modifications in the system of wage regulation. Starting with January 1976, it has been supplemented by a linkage of average wage growth to productivity growth. The normative linking average wages to productivity is set annually and can be made binding by the appropriate minister if enterprises do not observe it voluntarily. For enterprises which have not yet converted to the new system of planning and financing, the normative is binding.[54]

In June 1977 the government carried out some further modifications in the SWR. They do not apply to all organisations managed by the provisions of the 1973 reform but are confined to four branches of industry for the present and will gradually be extended to others.[55] The modified system[56] affects output added, disposable wage-bill and normatives. It unifies the criteria for the calculation of output added by associations in general, but in particular it makes it mandatory for each association to use uniform criteria for all the enterprises in its jurisdiction. The differences which have developed in the detailed conceptualisation of output added since 1973 have made it technically difficult for the centre to assess the performance of enterprises and have made the decision-making process difficult with regard to the economic levers to be applied to enterprises.[57] What is of no less importance is that the modified system tries to cope with the unjustified profits which result from unwarranted increases in prices. Such profits are to be excluded from output added through taxation,[58, 59] a policy which Hungary has applied for some time.[60]

To improve market equilibrium the disposable wage-bill is equipped with a built-in stabiliser additional to the deduction of unjustified profit. The stabiliser takes the form of a charge (*obciążenie*) on the increment of the disposable wage-bill, payable to the branch reserve fund[61] which is not a new institution. It came into being in 1974 in response to two disturbing phenomena in the new associations, which were already indicated: the considerable reserve funds which some had managed to accumulate and the unfavourable development in employment. In order to cope with this new situation, organisations were required to pay into the newly created branch reserve fund for increases in employment and increases in the disposable wage-bill above 4 per cent. The charges for the latter were set at a relatively low level.[62] What is new in the modified system[63] is that charges for wage increases have been made progressive, and the trigger threshold for paying charges has been made dependent on an annual agreement between the Planning Commission and individual branch ministries. With this provision the system of uniform charges—introduced in 1974—is no longer applicable to associations which have converted to the modified system.[64]

The main purpose of the branch reserve fund is to keep wage growth in associations in accordance with the state annual plan without eliminating the linkage of wage growth to output added. It is hoped that the obligation to pay a charge to the branch reserve fund in addition to the transfer of funds to the association's reserve fund will force wage restraint on enterprises. For this purpose the trigger threshold is set at a level corresponding to the state plan target for wage growth. Only wage increases above this threshold will be burdened by a charge.[65, 66]

These charges should not be confused with taxes. Charges, unlike taxes, remain at the disposal of the ministries and can be used with the approval of the Planning Commission for future increases in basic wages or for financing new employment.[67] With this statement we have touched upon the change which occurred in the concept of the reserve fund. Originally the idea of establishing reserve funds emerged as a solution to a possible shortage of funds for the required wage payments and/or managerial incentives. In the course of time the reserve funds have been changed into an instrument of wage control and increases in basic wages.

Finally, the normatives are fixed for two years only, 1977–8; for the remaining two years of the present five-year plan only estimates are set.[68] It is not yet clear why the central planners abandoned the original plan of setting long-term normatives for the whole five-year plan period. One of the main reasons seems to have been the reluctance of the authorities to commit themselves to normatives (for whose setting it is not easy to find objective criteria) in a situation in which wages grow fast.

The modifications of 1977 brought about a stricter Bank control over disbursement of wages. All enterprises and associations are obliged to submit data to the Bank on the planned disposable wage-bill, wage outlays, output added and normatives. It is the duty of the Bank to verify the figures and, if some irregularities are discovered, to report them to the supervisory bodies of enterprises. If the projected wage expenditures exceed the disposable wage-bill, an enterprise has to cover the difference from the 'anticipated loan'[69] or from its own reserve fund, and if these are not sufficient, then from the reserve fund of the association. If all these sources are not sufficient, the Bank will extend a repayable loan on which interest must be paid.[70] If an overdraft of the wage-bill occurs at the end of the year, the Bank is empowered to withhold funds from the bonus fund for managerial staff to the extent of the overdraft for which a loan is extended. There is only one exception: bonuses for foremen are not to be suspended.[71] It is not clear whether the bonuses of managers can be recovered after the overdraft is settled. If not, this would mean a real tightening of the sanctions for overdraft. According to the provisions of the reform of 1973, top managers could be penalised by having their bonus maximum reduced to 50 per cent. This penalty was to be meted out by the supervisory body.[72]

Nowadays Poland has three different systems of wage control: one, the old pre-reform; the second, that of the reform of 1973; and the third, the modified system of the 1973 reform. The pre-reform system is a direct system, similar to the Soviet one; the modified system is, in systemic terms, somewhere between the old system and the system of 1973 which we denoted as a mixed system. There is no doubt that the modified system means a step in the direction of centralisation of wage regulation and a strengthening of the administrative elements in the system. Associations applying this system are presently assigned by the centre not only the normative 'R', but also a scale of charges with a trigger threshold. 'R' is meant to be fixed for a certain length of time whereas the rest are fixed annually. Thus the greater certainty and predictability which the normative 'R' is supposed to bring to enterprises for the sake of extending their time horizon for decision-making is to a great extent undermined by the system of charges to the branch reserve fund. By the right given to ministries to determine the charges and the trigger threshold, their intermediary role and their administrative position in the system of management have been increased. This is not an accidental occurrence but the result of a more general change in policy. '... Parameters for WOG (associations) do not result directly'—writes K. Golinowski—from the substance of the central plan, but are retranslated on the level of the branch (ministry) in the language of the system WOG.'[73]

The new modifications give rise to some questions. For our purpose the most important one is: how do the authorities plan to keep wages within the limits of the plan if the trigger threshold on tax payments is to be set—according to K. Golinowski—at the level of the state plan target for wage growth? It should not be forgotten that the charges for wage increases are not exorbitant if the revenues of enterprises are considered.

9 Wage Regulation in Hungary

In this chapter we will concentrate on the SWR in Hungary since the economic reform of 1968. Up to 1968 Hungary relied on the SWR introduced in 1957, under which the centre assigned binding rates of wage increases to enterprises through the hierarchical channels of management. These averages were not allowed to be increased even if enterprises over-fulfilled the plan targets in productivity.[1] It was argued that there was no way to measure productivity adequately on the enterprise level. The real reason for this measure, however, was the determination of the authorities to hold down wage increases. Due to the political events of 1956 the control over wages got out of hand and nominal wages increased dramatically in 1956–7.

THE REFORM OF 1968

The reform of 1968 substantially increased the jurisdiction of enterprises in the planning of wages. Enterprises were given the right to plan their wage-bill, average wages and the number of employed. In the first three years (up to 1971) the growth of wages was closely tied to profit; increases in basic wages (as in bonuses) were financed from the bonus fund which was fed from profit. Since considerable fear of inflation was generated by uncertainty about the impact of the price reform effected in January 1968 and about how the new regulators would work, the government set a ceiling on basic wage increases at 4 per cent. The ceiling was supposed to be temporary but later it was extended up to 1971.[2]

Profit, defined as the difference between enterprises' revenues and their production costs (increased by a charge on fixed and working capital, a contribution to social insurance and a tax on wages),[3] has been chosen as the evaluation indicator and incentive. According to the provisions of the reform, profit was divided into two parts: one, earmarked for complementing wages (the sharing fund), and the other to be used for the development of enterprises, the development fund. The first part was taxed progressively, the second linearly.

This division of profit into two parts proceeded according to the following formulas:

$$P_w = \frac{W}{\alpha W + K}$$

$$P_d = P - P_w \quad \text{or}$$

$$P_d = \frac{K}{\alpha W + K} \cdot P$$

It follows that:

$$\frac{P_w}{P} = \frac{W}{\alpha W + K}$$

$$\frac{P_d}{P} = \frac{K}{\alpha W + K}$$

where P_w is the sharing fund (subject to taxation);
P_d the development fund (subject to taxation);
P profit;
α the multiplier;
W the wage-bill and
K the fixed and working capital.

Profit before taxation was thus divided up according to the enterprise's capital—wage costs ratio. In order to increase the portion of the sharing fund, a wage-bill multiplier was introduced. It was set on the average at two and differentiated according to sectors of the economy in order to mitigate the consequences of the great differences in K/W.[4] The portion of profit assigned to the sharing fund was, as already mentioned, taxed progressively. In the first year of the reform the part of the taxable sharing fund which corresponded to 3 per cent of the wage-bill was tax exempt. In each of the two following years the percentage of the portion exempt from taxes grew by 2 per cent. The remainder of the sharing fund was taxed at a rate of 20–70 per cent. From the sharing fund thus created 10 per cent was earmarked for the reserve fund. The balance could be used for distribution.[5] The sharing fund was used for financing wage increases, not only current increases but also the cumulated increments in basic wages from the beginning of the reform, as well as for bonuses, year-end rewards, the subsidising of housing and social and cultural purposes.

In 1968 money incentives were distributed in three parts. The first part, for top managers, could amount to 80 per cent, and the second, for middle cadres (heads of departments, supervisors, engineers, etc.) to 50 per cent of their total basic salaries as a group, and the third, for blue-collar workers and others, to 15 per cent of total earnings.[6] This relationship, 80 : 50 : 15, was to be retained even if the sharing fund was too small to ensure for every group the mentioned amount of bonuses in relation to their salaries and wages.

The planners soon discovered that the new incentive system could not work as expected. The reform left in the hands of the managers, in cooperation with trade union organisations, the distribution of the sharing fund, including the option of using it either for increases in wages or for bonuses. However, the interests of managers—mainly top managers—and workers were conflicting. Top managers were interested in maximising profits and in using the sharing fund for bonuses rather than for basic salary increases. From an increase in bonuses corresponding to 1 per cent of the wage-bill they themselves could expect an increase equivalent to an increase in salaries of 4–5 per cent. There was also another reason for their preference for bonuses: increases in basic salaries had to be approved by authorities above the level of enterprises. On the other hand, wage earners were not nearly so interested in bonuses; an amount of bonuses corresponding to 1 per cent of the wage-bill meant for them an increase of 0.7–0.8 per cent in total average wages.[7]

The conflict between management and workers was also enlarged for another reason. The great differentiation in bonuses was a deliberate policy aimed at widening wage differentials in order to stimulate the maximisation of profit and higher efficiency. One way of maximising proft is at the expense of wages. There are indications that top managers resorted to this method, in some cases by increasing performance norms. Workers responded to such pressures with the only possible weapon in conditions where strikes are not allowed—a slowdown.[8] Wage earners' resentment at the system of distribution of bonuses cannot be explained only by material interest. Undoubtedly social considerations also played an important role; wage earners regarded the system as social discrimination against them. For understandable political reasons the Hungarian authorities abolished the system in 1969.[9]

The SWR did prove relatively effective in keeping wage increases within the limits envisaged by the planners. In the period 1968–70 average wages (inclusive of bonuses) increased by 15 per cent.[10] Considering that the limit for basic wages was 4 per cent and that enterprises, after long and strict control over wages, had once again some say about their growth, the result seems to have been satisfactory.

Many enterprises, however, expressed great dissatisfaction with the system. The provision that increases in basic wages from the start of the reform should be financed from the sharing fund became for many enterprises something beyond their financial capability. In addition it reduced the size of funds which could be used for bonuses. Therefore in 1970 the system was changed somewhat; only 70 per cent of the wage increases in 1970 were financed from the sharing fund.[11]

Contrary to the planners' intentions, the SWR (based on average wage regulation) combined with a wage ceiling turned out to be a stimulus for hoarding labour. Thus the original fear that the new system might bring about a reversal in the usual approach by enterprises to utilisation of labour and thus produce unemployment,[12] turned out to be unfounded. In looking for ways to circumvent the ceiling for average wage increases, enterprises found a further source for financing increases in average wages in savings made by hiring less skilled workers who could be paid wages below the average.[13,14] (A direct increase in average wages—as already noted—had to be financed from the sharing fund.) Hiring new workers also had the advantage that they were not eligible for bonuses for some time.[15] Hoarding of labour was just one side of the coin; on the other side were shortages of labour in many branches and enterprises. A slowdown in productivity growth was the final effect.[16]

MODIFICATIONS IN 1971

The five-year plan for 1971–5 brought about changes in the SWR and in incentives. The wage ceiling and the financing of wage increases from the sharing fund were dropped. Instead, wage growth was linked to gross income (profit + wage-bill) per employee of the enterprise. For every 1 per cent increase in gross income per employee over the previous year, enterprises were allowed to increase average wages by 0.3 per cent. The costs of wage increases had to be included in production costs.

Had the reform not introduced a new tax to regulate basic wages, the dependence of the latter on profit would have been weakened due to the new provisions. It would only have been sustained through gross income being the indicator. However, what is perhaps more important, basic wages would have escaped the indirect regulative influence of taxes, which have been used since 1968 to regulate the sharing fund which is fed from profit. The new tax was paid (and is still paid in a modified form) from the sharing fund, thus strengthening the indirect dependence of wages on profit. The use of taxation as a regulator of wages has become important also because the planners wish to give enterprises certain choices as to the distribution of funds at their disposal. Enterprises can use the sharing fund for taxes on wage increases above the limits entitled by performance or for bonuses.

Enterprises paid a single tax (50 per cent of the additional wage costs) as long as wage increases were confined to a percentage increase corresponding to the achievement in gross income per employee. For wage increases above the limit they had to pay a progressive tax ranging from 150–400 per cent of the additional wage costs.[17] In addition, changes were made in the taxation of the part of profit earmarked for the sharing fund. The taxation remained as before linked to the wage-bill, but average tax rates were increased and the progressivity was decreased. On the other hand, the wage multiplier was increased from an average of two to three, which meant that enterprises received a greater portion of profit for the sharing fund.[18]

Many expectations were connected with the new changes in the wage regulation system. Planners believed that they had found the right cure for the problem of labour hoarding. Is it not reasonable to expect that managers would care more about economising on labour if increases in wages depended on productivity? It is obvious that the same gross income produced by fewer people means a greater rate of increase in gross income per employee and more funds for wage increases with more favourable tax conditions.

The change in the wage multiplier and the elimination of the wage ceiling were also supposed to work in this direction. A more generous allocation of funds to the sharing fund had not only to ensure funds for paying higher taxes on that part of profit which was earmarked for the sharing fund, but it also had to make available funds for needed taxes on wage increases, thus enabling enterprises to increase average wages without having to resort to hiring lower paid workers. Apparently confident in the effectiveness of the new measure, the government abolished the 1970 tax which had been intended to cope with the hoarding of labour.[19] The planners also hoped that the fact that the new indicator for wage growth being the same on the enterprise level what national income per capita is on the macro-economic level would enable a better regulation of wages in accordance with the growth of national income and would have a stabilising effect on the economy.[20] Finally, the changes were also intended to enable a greater differentiation in wages as a stimulus to higher efficiency.[21] The above-mentioned reduction in tax progressivity was aimed precisely at this objective.

The modifications in the SWR eased the problem of labour hoarding to a certain degree but widened the wage differentials to such an extent that it became a political embarrassment to the government.[22] The close linkage of wage growth to profit was very much blamed for this situation, which was all the more objectionable because the differences in profit between enterprises in relation to the wage-bill are also a result of factors which are beyond the control of enterprises. In particular, intrabranch and interbranch differentials developed in an undesired way; e.g. wage increases in coal and energy production remained far behind increases in other branches.[23] The new system of wage rates[24] should be blamed to a certain degree for the unfavourable development of intra-enterprise differentials.[25]

The heavy taxation on wage increases for the purpose of shielding the economy against wage inflation turned out to be an insurmountable obstacle to reasonable wage increases for many enterprises.

Faced with these problems, the government resorted to a compromise which perhaps was the best solution under existing conditions. In order to correct the situation some direct measures were applied without dismantling the indirect system very much. To placate the workers, the government ordered a wage increase (effective in March 1973) for wage earners in most of the state enterprises which had partially to finance the increase.[26] Later on, other groups received a raise. Needless to say, this was a deviation from the policy established during the reform, according to which regulation of wages was to be enforced primarily through indirect methods. Some branches whose products are sold at fixed prices (such as coal and energy production) and which fared poorly under the existing system became subject to a direct assignment of the wage-bill by the centre. Enterprises which were not able to increase wages adequately because they failed to produce enough profit were granted a tax exemption.[27]

MODIFICATIONS OF 1976

The experience with the wage regulation system after 1971 and the changes applied in this period set the stage for new modifications introduced in January 1976 at the start of the new five-year plan. One of the motives for the modifications, which also influenced their direction, was the desire to pass on to the consumers, at least partially, the consequences of the worsening terms of trade. For this purpose the planners planned much lower rates of increases in real incomes and real wages. Since the central authorities do not set wage-bills and bonus funds to most enterprises directly, they imposed higher taxes on enterprise's profits earmarked for bonus funds[28] and reduced normatives for average wage regulation. Yet for stimulative reasons they eased taxes on wage increases up to certain limits anticipating higher price increases anyway. First a general characterisation of the changes will be given, followed by a more detailed explanation.

1. The new changes mean a move away from uniformity, and they represent an effort to adjust wage regulation more to concrete conditions. The changes are not confined to normatives, but also affect the system itself. Apparently planners believe that this arrangement will lead enterprises to more self-sufficiency in matters of finance and will reduce the need to deal with enterprises' requests for exemption, preferential treatment, etc.[29] Does this mean that the principle of uniformity has not proved practical? There is no doubt that this principle cannot be equally just to all. It is reasonable to assume that the same conditions will never obtain in all enterprises; whatever rules are applied, some will benefit and others will be at a disadvantage. This shortcoming may be offset by the advantages which uniformity can provide. However, in Hungary the relatively distorted price system, the four different methods of price formation and the different level of technology in enterprises make 'competition' quite unequal under conditions of uniformity. In addition the imposition of heavy taxation on wage increases put many enterprises, particularly before 1976, in a difficult financial situation which required government interference, thus reducing the usefulness of the principle of uniformity.

As is known, there are also institutional constraints on the application of uniformity; socialist countries—and this is just as true of Hungary—are reluctant to let an enterprise go bankrupt because of inefficiency. Heavy taxation and distorted prices make the necessity of bailing out laggard enterprises more frequent.

2. The changes of 1976 mean a loosening of the linkage of wage growth to profit.[30] The Hungarian government has been torn between

two principles: to make the system more stimulative, and to keep wages within the limits of the plan and avoid great wage differentials. (In practice, the problem lies not so much in average wages as in wage differentials.) Sometimes the first principle prevailed, as during the reform of 1968 and 1971. When the effects of stressing greater motivation showed up in practical measures and wage differentials started to widen, political dissatisfaction built up and the government was forced to retreat, something which happened in 1969 and 1973. The modifications of 1976 again indicate a retreat.

3. With this we have arrived at a further characteristic, which is that the government has reserved the right to interfere directly with wage differentials (as it did in 1973) whenever it feels that they do not develop in accordance with its objectives.[31] The government also intends to use this right for wage adjustments in branches which lag far behind the average level (textile, trade). This provision shows that the government has come to the conclusion that it is very difficult to achieve by indirect methods a correct balance between the above-mentioned principles of wage policy.

4. Direct methods of wage regulation have been substantially expanded, of course, at the expense of indirect methods.

5. Finally the modifications are marked by an innovation, namely, by the introduction of a guaranteed wage increase to be changed annually as a partial compensation for the expected rate of inflation.

On the whole, the direction of the changes is towards more government interference and controls,[32] though in some respects the system has been made more flexible. Viewed from a systemic angle, it is possible to argue that Hungary has two systems nowadays: an indirect, with some qualifications, and a direct one. But both systems are applied in two forms, and therefore Hungarian economists usually talk about four methods of wage regulation.[33] Though the four methods are not all equally represented, A. Timár stresses that no method should be regarded as preferable; all should be treated on an equal footing.[34] First, the indirect system will be examined which consists of the old system of average wage regulation with some changes and of regulation of the wage-bill.

Linkage of Average Wage Regulation to Performance

Average wage growth is linked as before to gross income per employee. While at the time of its introduction this linkage was applied in the vast majority of the state enterprises, the modified system of 1976 substantially reduced its role. According to A. Timár enterprises which adopted this system employed 34 per cent of the labour force in the *khozraschet* sphere.[35] (Computed by number of enterprises, the weight of this linkage was much higher since it was used in many smaller enterprises.[36]) As will be shown later, its weight was further reduced in 1978.

The linkage of average wage growth to gross income per employee was applied in most of the engineering industry, chemical industry, half of light industry and in most of trade,[37] in branches where the initiative of enterprises may be a source of higher efficiency, where economic growth from the societal viewpoint is desirable and where growth is accompanied by changes in output mix.[38] In contrast to the pre-1976 situation when the normative for a 1 per cent increase was 0.3 universally, it is now differentiated. In most enterprises it is 0.25. In addition to this wage increase based on performance, enterprises are entitled to an additional raise, an annually fixed allowance for expected inflation. For 1976 this allowance was fixed at the rate of 1.5 per cent of the average wage.

In all profit making enterprises, regardless of their wage regulation methods, the division of profit into two parts according to a formula set by law has been abolished. According to some authors the change was necessary, since the old arrangement hampered incentives in enterprises with a fast growing capital; in such enterprises even a considerable increase in profit meant only a small allocation of funds to the sharing fund.[39] It seems that a more important reason was the desire of the authorities to reduce the flow of funds to the bonus fund.

On the one hand, the planners have granted enterprises the right to divide up profit according to their own priorities. On the other hand, they have made sure by a series of provisions that enterprises will not be able to feed the sharing fund excessively nor use it for purposes the planners considered inappropriate. A heavy progressive tax is levied on the part of profit earmarked for the sharing fund,[40] whereas funds going into the development fund are exempt from taxation. In addition enterprises are obliged to use the sharing fund in a certain order.[41]

Guaranteed increases in wages, as well as increases in wages corresponding to the increase in gross income per employee, are not subject to taxation as long as both together do not exceed the set 6 per cent limit. 'Deserved' wage increases above the 6 per cent are subjected to a proportional tax, whereas all wage increases above the limits of performance are progressively taxed. Both taxes are payable from the sharing fund. The new tax arrangement means that the linkage of wage growth to profit is considerably loosened. First, the guaranteed wage allowance is not linked to any performance. Secondly, in contrast to the pre-1976 period, enterprises no longer pay taxes on wage increases up to 6 per cent resulting from performance. The planners have, however, no reason to fear that wage increases will be inflationary. As already mentioned, wage increases above the limits are burdened with heavy taxes, and in addition the sharing fund from which taxes are to be paid is also limited due to taxation.[42]

Linkage of Wage-Bill Regulation to Performance

This is not an entirely new method of wage regulation; it was already applied in 1968 in state agriculture and forestry. In 1970–1 it was extended to some branches of the food industry. In 1972 the central authorities started to experiment with regulation of the wage-bill (instead of average wages) in some selected industrial and construction enterprises. On the one hand, the reason for this experiment was that most enterprises which had already used this method had apparently achieved good results. On the other hand, because dissatisfaction had built up with the prevailing system, regulation of average wages linked to gross income per employee had not brought about as great a reduction in labour hoarding as had been hoped for. According to P. Bánki, in 1973 most enterprises still had labour reserves between 5 and 20 per cent,[43] a situation which, in light of the growing labour shortages due to an inefficient labour utilisation, was viewed by planners with concern. It seems that the conclusions drawn from the experiment were not as optimistic as was originally expected; some experimenting enterprises refused to adopt it. Regulation of the wage-bill can be advantageous to enterprises which have labour reserves, but once they are exhausted or close to exhaustion, the advantages vanish. On the other hand, this system may produce undesired changes in wage differentials, something which the planners try to avoid.[44]

In 1976 wage-bill regulation was applied, apart from sectors in which it was already practiced, in part of the engineering industry (in some of the biggest trusts), in most of the construction industry and in some other enterprises (altogether in enterprises with 35 per cent of the work force).[45,46] It was applied primarily in sectors where expansion of output can be achieved by increases in productivity. Growth of the wage-bill is nowadays predominantly linked to value added, which is defined as a total sum of wage costs, profit, charges on capital, bank expenses and depreciation.[47] The normative is differentiated: in 1976 it was one in state farms (1 per cent increase in output was rewarded by a 1 per cent increase in the wage-bill) and in industrial enterprises not higher than 0.4.[48]

Wage-bill regulation is not a pure method; it is combined with average wage regulation which serves as a built-in brake. If enterprises wish to increase average wages above 6 per cent (computed on the basis of fully employed) they must pay a proportional or a progressive tax depending on their performance. However, if the 6 per cent is not 'deserved' by performance (inclusive guaranteed rate), they must first pay a progressive tax on the balance. In order to encourage enterprises to labour saving, the tax progressivity is much less steep than in the case of average wage regulation.[49]

Direct regulation of wages is also applied in two forms: through the wage-bill and through average wage regulation.

Direct Average Wage Regulation

This method is applied in coal and electricity production, in some branches of the food industry, in part of transportation and in part of foreign trade and services.[50] It is applied in economic organisations where the conditions of economic activities are determined from above, with the result that profit cannot be expected to grow sufficiently to ensure a reasonable growth of wages. In addition it is applied in services where an increase in employment is desirable.[51]

The application of this method means that the centre fixes the rate of increase in wages. Hungary had already applied such a system before the reform of 1968. The present system differs in one respect—in that enterprises are given the right to increase wages above the limit, provided they are willing to pay a heavy progressive tax.[52] It seems that the rate of wage increase is fixed annually. For 1976 it was fixed at the level of $4\frac{1}{2}$ per cent, including the guaranteed allowance.

Direct Wage-Bill Regulation

This method is applied in research institutes, organisations for investment planning and in some branches of transportation. Because the centre is not interested in the expansion of employment there—in some cases it even wants to reduce it—these organisations, while having features in common with the previously discussed organisations, are nonetheless treated differently. As in the former case, the centre fixes the rate for wage-bill increase. For 1976 it was fixed at the rate of $4\frac{1}{2}$ per cent (inclusive of guaranteed allowance). If an enterprise reduces its labour force, it can increase average wages up to 6 per cent without payment of taxes. Increases above this limit are subject to the same tax rates as regulation of the wage-bill on the basis of performance.[53]

Economists disagree about how to evaluate the modifications of 1976, as the recent debate in *Közgazdasági Szemle* has shown. For example one author maintains that they will reduce the stimulative power of the system and encourage enterprises to expand the bonus fund at the expense of the development fund.[54] Others argue that bonuses will grow to a greater extent from the basic wage-bill (from which bonuses are also payable) and that higher taxes on the part of profit going to the sharing fund will benefit the development fund.[55]

From the limited information available, it is clear that the bonus fund in 1976 declined and bonuses increased at a much slower rate than in 1975 and than anticipated.[56] On the other hand, basic average wages, though they too lagged behind the 1975 growth rate, grew faster than planners had assumed they would.[57] In 1976, 30 per cent of the work force achieved more than a 6 per cent increase in wages. This means that even the tax was not an insurmountable hurdle for wage increases. In 1977 the number of workers whose wages had risen more than 6 per cent was even greater; the average wage increase was 6 per cent against 5.7 per cent in 1976.[58, 59]

There were several factors which enabled higher wage increases than expected. Most enterprises managed to accumulate some wage reserves during the years and even increased them during 1976.[60] (Part of these funds was used for bonuses.) Enterprises with wage-bill regulation not only achieved higher wage reserves but also benefited more from the possibility of increasing average wages through labour saving. In these enterprises wages grew faster than in enterprises with average wage regulation.[61]

Basic wages could grow faster than assumed also because of the faster than expected growth of profit, and consequently both wage growth regulators also grew faster than was anticipated.[62] This also meant that enterprises had more funds available for taxation purposes in case they increased wages above the performance limit. Naturally not all the profit produced was the result of increased efficiency.[63] Finally, the fact that consumer prices grew faster than anticipated had some indirect impact on wage increases.

Taxation played a greater role than was foreseen. Without taxation, the factors mentioned above would have pushed up wages much above the actual growth. In addition, it also dampened tendencies to wider wage differentiation. Hence taxation proved an effective barrier to undesirable wage increases and differentials.

In January 1978 the government again carried out some changes. Wage-bill regulation has been expanded at the expense of average wage regulation not only in the sphere of the indirect system but also to some extent in the direct system; thus it has become the most important method of wage control.[64] It seems that the main motive for this change lay in productivity considerations, in the fact that productivity grew faster in enterprises with wage-bill regulation than in enterprises with average wage control.[65] Some authors are doubtful whether the implication that the method of wage control was instrumental in promoting productivity is valid. Rather they suggest labour shortages played a more important role in enterprises' effort to increase productivity.[66] It also seems that the experience with wage-bill regulation has been favourable because it has been applied in enterprises where conditions have been conducive to increases in productivity. In this connection also, more advantageous taxation for wage increases than in enterprises with average wage regulation may have played a role. If not for these two factors and particularly the first one, there is no explanation why enterprises with average wage regulation under Hungarian conditions (where the indicator is a productivity indicator) could not have achieved the same result.[67]

The experience of the last two years has also showed that the normatives were too high in some enterprises with wage-bill regulation and therefore they were reduced.[68] Otherwise the system has been retained, including the 6 per cent as the trigger threshold for tax payments.

10 Wage Regulation in Czechoslovakia

Up to 1958 Czechoslovakia applied the classical Soviet SWR apart from a departure from the Soviet practice in one respect. In 1956 it introduced a change in the relationship between the over-fulfilment of output targets and the growth of the wage-bill. Until then, the wage-bill of enterprises could increase proportionally to the rate of over-fulfilment of the gross value of output target. This arrangement was—as already noted—prone to inflation. When in 1956 the planned global wage fund was overdrawn, the government decided to impose a more rigorous control; one of the most important measures was to introduce adjustment coefficients lower than unity for the over-fulfilment of plans.[1]

THE REFORM OF 1958

In 1958 the first attempt was made to reform the administrative system of management by granting enterprises greater decision-making power. Changes in wage regulation and in the incentive system were an integral part of this attempt. The central authorities ceased to assign the wage-bill or average wages directly to enterprises; instead, wage growth was to result from the assigned minimum growth rate of productivity and a long-term normative for the growth of wages for 1 per cent rise in productivity. If enterprises reached the minimum rate of productivity growth, the average wage increased according to the normative. For productivity growth above the plan, average wages could grow by a higher normative. Under-fulfilment of targets was penalised by a higher normative than used for over-fulfilment.[2] The incentive fund also underwent a change; it became linked to increment in profit (instead of to output indicators) and a long-term normative. The 'reform' was short-lived; it was brought to an end by the developing economic crisis which to a great degree resulted from the ambitious and unrealistic third five-year plan for 1961–5. In 1962 Czechoslovakia returned to the old system.

It is of interest for the purpose of this study that the 'reform' represented the first attempt to use long-term normatives for regulation of wages and bonuses. The idea itself did not originate in Czechoslovakia; it emerged first in a debate by Soviet economists in the columns of the *Kommunist* in 1956.[3] The introduction of long-term normatives ended in complete failure. No doubt, the aim pursued by the adoption of this tool of wage control, namely, to give some certainty to enterprises with regard to the evolution of wages and bonuses, was a step in the right direction. But putting normatives into operation in an administrative system—which Czechoslovakia adhered to despite the 'reform'—turned out to be almost an impossibility, as could be anticipated. The methodological instructions for the introduction of long-term normatives of 1958 contained a clause stating that normatives were set on the assumption of a certain rate of economic growth during the five-year plan and would be changed if the actual growth deviated significantly from the planned.[4] When this soon happened, the planners were confronted with such a tremendous task that they were forced to capitulate, and the long-term normatives were in fact turned into annual ones. The former were in force for such a short time that they could not really affect the behaviour of managers in the desired way.[5,6]

THE REFORM OF 1966

In 1966 Czechoslovakia embarked on a far-reaching reform of the system of management. The reform aimed at changing the relationship between the central authorities and enterprises, at changing enterprises from a position of a quasi-subordinated board to units of relative economic autonomy. To achieve this, annual plans with centrally imposed binding targets were basically eliminated. Instead, elements of the market mechanism were expected to push enterprises to rational behaviour. The performance of enterprises was no longer evaluated on the basis of the extent to which plan targets were fulfilled; instead, the degree of economic efficiency achieved in the process of satisfying demand was supposed to be the yardstick.

Naturally the system of wage regulation and the incentive system had to be adjusted to the principles of the new system of management which have only been briefly sketched here. Commencing in 1966, central authorities ceased to assign wage-bill and average wages to enterprises. Managers of enterprises were given the right to make decisions on matters of remuneration and the number of personnel, always of course within the framework of valid legal regulations. In addition they were allowed, in cooperation with trade union organisations, to determine wage forms and the methods of their application, as well as incentive schemes and the distribution of bonuses.

In the new system, funds for wage payment were no longer to be allotted to enterprises; instead, enterprises had to 'earn' them by their own economic activities. This meant that the personnel had to take over responsibility for the performance of their enterprise. Of course this responsibility could only be partial since full responsibility would necessarily lead to great differentiation in employment incomes—largely without taking account of the merits or faults of the personnel.[7] The principle of partial responsibility brought about two new phenomena in remuneration: first, a share in the annual net returns (year-end rewards), and secondly, a wage guarantee.[8] The purpose of introducing a sharing in annual returns (termed, in official parlance, reward for collective results) was to arouse the interest of all the personnel in the overall performance of the enterprise. The institution of a wage guarantee[9] aimed at protecting the personnel from income losses due to the poor performance of the enterprise for whatever reason. This was also important in order to allay fears that the new system might harm the interest of the workers.

The new management system encompasses two incentive systems based on two different evaluation indicators. In enterprises where performance was evaluated in terms of gross income[10] (which was the case in the vast majority of enterprises), taxes were levied on gross income. In a minority of enterprises, profit was the evaluation indicator and taxes were levied on it. The architects of the reform regarded as essential—as already noted—that taxes be levied on the magnitude which serves as an evaluation indicator.

In enterprises where gross income was the evaluation indicator, the wage-bill[11] was the portion of gross income which remained with enterprises after they had carried out their obligations to the State in the form of taxes[12] and to the banks, and after the obligatory formation and replenishment of funds. The wage-bill so formed was used for basic wages and also for bonuses. Enterprises were given the right to determine whether they would use the increment in the wage-bill for basic wages or for bonuses.

In enterprises where the material interest was tied to profit, the wage-bill consisted of two funds. The basic wage-bill, earmarked for basic wages, continued to be fixed in an administrative way since the reformers believed that if profit was the evaluation indicator, the basic wage-bill would have to be determined from the centre. The other fund, the so-called remuneration fund, served for bonuses and year-end rewards and was that part of profit which remained with the enterprises after they had fulfilled obligations which were the same as in enterprises with gross income as indicator.[13]

The Czechoslovak reformers were the first in the Soviet bloc to introduce regulation of wages and bonuses by taxation. However, during a short period of three years the method of taxation was changed twice. What is no less important is that the government supplemented it for some time with an administrative method. As will be shown, the method of taxation which was applied was not sufficiently well thought out to be the proper solution to the problems it was intended for. In a sense this is not surprising when one realises that there was little experience with this kind of regulation, and in addition there was not enough time to take account of all aspects involved.

The first attempt to regulate wages by indirect methods was made in 1966 when the so-called supplementary tax was introduced. It was a tax on the wage-bill and bonus funds (the latter for those enterprises with an incentive system based on profit), and thus indirectly on gross income and profit respectively for wage increases which exceeded the planned average wage by a certain percentage. In 1966, the first year of its imposition, average wage increases up to 6 per cent were exempt from tax: in 1967, if the tax had not been abolished, the cumulative exemption (including that of 1966) would have been 10 per cent and in 1968 14 per cent. For wage increases above the exempted limits enterprises had to pay a progressive tax.[14] The progressivity of the tax rates was intended to discourage enterprises from an excessive increase in their wage-bills and thus to hold down growth of average wages to an acceptable rate. It is worth noting that the tax exempt wage growth rates were fixed higher than wage increases had been previously. To some degree the concept of the supplementary tax reflected a new approach to the growth of wages.

This tax was supposed to be in force for at least three years, but a year after its introduction it was replaced by the stabilisation tax. The main reason for this change was the decision to introduce a uniform tax on gross income instead of the existing differentiated tax, a provision which was part of a comprehensive package aimed at speeding up the establishment of the new system of management. It was believed that a uniform tax would put an end to the practices of the administrative system which benefited poorly performing at the expense of efficient enterprises.[15] And the new stabilisation tax, constructed according to the new concept of taxation,[16] was supposed to strengthen this intention. Its adoption was also motivated by the desire to regulate employment together with wages.

The new tax was levied on the enterprise's actual (not planned) wage-bill regardless of the incentive system (that is, whether it was based on gross income or profit). Since the amount of the total wage-bill is influenced by the movement of average wages and the number of employees, the stabilisation tax affected both these factors.

The formula for the stabilisation tax in 1967[17] was the following:

$$T = \frac{30}{100}\left(W - \frac{90}{100}wE\right) + \frac{r \cdot W}{100}$$

where W is the actual wage-bill in 1967;

w the planned average wage in 1966;

E is number of employed in 1967 and

r is the growth rate of employment in 1967.[18]

As the formula shows, the tax consisted of two divisions. The first on the left, which aimed at regulation of average wages, was uniformly applied. By contrast, the second division, which was supposed to control employment growth, included many exemptions.[19]

The introduction of the stabilisation tax pursued three main goals:

1. Anti-inflationary—to discourage enterprises from excessive allocation of funds to the total wage-bill and thus to keep average wages in line with the planned distribution of national income. For this purpose the government also stipulated that the wage-bill should grow more slowly than the allocation for investment.[20]

2. To discourage enterprises from increasing output by an increase in employment. It was expected that the increase in labour costs would encourage enterprises to substitute capital for labour and thus increase productivity.

3. To have a built-in stabiliser—hence the name 'stabilisation tax'—which would keep wage differentials between the 'material' and 'non-material' spheres within acceptable limits. With the increase in average wages in the 'material' sphere (where only the stabilisation tax was

collected) the yield from the tax would increase, and the government would thus get funds for increasing wages in the non-material sphere.[21]

It soon became clear that the stabilisation tax alone could not effectively regulate wages according to the plan. In 1967 Czechoslovakia carried out a wholesale price reform which increased prices by 29 per cent instead of the expected 19 per cent.[22] This put the gross income of enterprises in a much more favourable situation than had been expected when tax rates were fixed on gross income. The situation was all the more complicated in that the price reform had brought about a considerable growth in wholesale prices, precisely in those branches of industry where average wages were already rather high[23] and where, from the point of view of society, there was no particular interest in keeping up the existing intensive investment drive.

The government was thus confronted with the problem of how to handle the new income situation which involved the inherent risk of a considerable increase in wages—not covered by an adequate rise in productivity and output—as well as in investments, which could set an inflationary spiral in motion. Perhaps the problem could have been solved by resorting to a differentiated increase in gross income tax, but the government feared that the abandonment of the uniform tax rate immediately after its introduction might undermine confidence in the new system of management. Therefore the tax was retained in the old form but supplemented by linking growth of wages to labour productivity. On the level of industry, an overall 0.68 per cent increase in average wages per 1 per cent increment in productivity was established[24] and differentiated by individual branches of industry and by enterprises.

The objectives of the newly introduced wage regulator were in substance the same as those of the stabilisation tax. It was expected that it would prevent excessive increases in average wages and would also discourage enterprises from increasing employment by giving them the prospect of increased wages for increased productivity. Unlike the stabilisation tax, however, the emphasis of the productivity indicator was on curbing wage increases. However, the stabilisation tax and the productivity indicator constituted different methods of achieving the above-mentioned objectives. While the former, an indirect method, was a product of the economic philosophy of the new system of management, the latter, mainly because of the way it was carried out, was a direct method and as such not a weapon from the arsenal of the new system.

Reference has already been made to the fact that the first division of the stabilisation tax had only a single rate which was applied uniformly to all branches of industry. It seems that the main reason for this uniformity was the desire to avoid favouring some enterprises. Such an approach could have been justifiable if the great differentiation in revenues among enterprises had been only the result of economic performance. Due to the fact that it was to a great extent a result of the unsuccessful wholesale price reform, as well as of the monopoly power of many enterprises, the stabilisation tax necessarily discriminated against enterprises which gained less from the wholesale price reform and/or which were not in a monopoly position. Had the tax been progressive, it could have eliminated many disparities that had arisen, mainly due to the wholesale price reform. The productivity indicator, precisely because of its differentiation, was more effective as a tool for regulating wages than the stabilisation tax.[25]

As of January 1969 the Czechoslovak government replaced the productivity indicator by a progressive tax on wage increases above 5 per cent. The new tax, which represented a further step towards dismantling the directive elements in the system of management,[26] resembled in its structure the above-mentioned supplementary tax and was in fact a surtax on the first division of the stabilisation tax.[27] If the average wage increase ranged between 5 and 7 per cent the tax was:

$$T = \Delta W(a - \tfrac{5}{100})20$$

where ΔW is the increment in the wage-bill and
  a the annual growth rate of wages (expressed in per cent).

If average wages increased above 7 per cent, the enterprise had to pay a further tax equal to the increment in the wage-bill.[28]

THE PRESENT SYSTEM

Not long after Dubček's leadership was toppled, a general attack was launched against the most important aspects of the reform. The new leadership under Husák embarked on a gradual elimination of elements of the market mechanism and on a revival of the administrative system of management in the name of promoting the authority of the state plan, the most important change being the reintroduction of annual plans with their binding targets. Naturally, this process of recentralisation could not but also affect the system of wage regulation and the incentive system.

In 1970 a new system of wage regulation was introduced which is basically still in force. According to it wage-bill formation is no longer in the hands of enterprises. Instead the central authorities assign to supervisory bodies of enterprises (in large-scale industry these are branch ministries) mandatory limits for the wage-bill[29] and the bonus fund.[30] The wage-bill is assigned as a fixed share in the results of economic activities which are predominantly expressed in marketed output.[31, 32] (The bonus fund is assigned as a fixed share of produced profit.) The supervisory agencies assign in turn the wage-bill to individual enterprises in their jurisdiction. Of course the normatives which determine the share of the wage-bill in marketed output are differentiated. If there are good reasons for doing so, they may also use another indicator or indicators.[33] According to recent regulations, effective since 1976, the supervisory bodies are also allowed to fix the share of wages in marketed output compared to the previous year.[34] It is not clear to what degree they take advantage of this. In cases where fixing the wage-bill as a proportion of marketed output is not warranted (e.g. because it is difficult to assess the growth of marketed output) it is possible today, just as it was in the past, to set the wage-bill as an absolute sum.

Over-fulfilment of planned targets in marketed output was rewarded by additional allocations to the wage-bill on the basis of an adjustment coefficient which was smaller than unity. According to the new regulations starting in 1976, enterprises are assigned at least two coefficients, a higher one for an over-fulfilment to a certain limit (expressed as a percentage) and a lower one for over-fulfilment above this limit. The system of two (or more) coefficients is aimed above all at dampening the growth rate of the wage-bill and thus represents an anti-inflationary measure. In fixing the exact value of the coefficients the following factors are taken into consideration: the share of wage costs in marketed output, the expected gains in productivity and the centre's interest in the expansion of the output of the enterprise in question.[35]

Supervisory bodies (similarly enterprises) can create reserve funds from funds allotted to them for wage purposes, which can be used for additional allocations to the wage-bill in case of changes in the plan and/or to cover the wage-bill overdrafts.

From the description given above it is clear that Czechoslovakia has basically returned to the old pre-reform system of wage-bill determination and regulation. There is however one difference: gross value of output has mostly been replaced by marketed output, which differs from the former in one important respect—it is a sales indicator. But it is not a genuine one conceived in rigorous terms; it includes not only the value of products which are sold but in most cases also the value of unfinished products and changes in inventories. It is doubtful whether, in practice, the difference between marketed output and gross value of output has important repercussions on the behaviour of enterprises, that is, whether in contrast to gross value of output, it encourages managers to have greater regard for consumer demand.

What is clear is that marketed output has many of the disadvantages of gross value of output. It encourages a shift in the output mix to material intensive products at the cost of labour intensive. The higher the value of materials embodied in products, the higher the value of marketed output. Thus marketed output indirectly encourages a waste of materials and energy. It also makes it advantageous for enterprises to buy services from other enterprises rather than produce them themselves. Thus it stimulates an extension of cooperation beyond economic rationality.[36]

One could argue that since such behaviour adversely affects the size of profits and contravenes the enterprise personnel's interest in profit, it will be corrected. This would be true if bonuses from the bonus fund, whose size depends on the amount of produced profit, played an important role in employment incomes, particularly of managers. Apart from top managers, the bonuses from the bonus fund for other managers are, on the average, relatively small (see Chapter 4).

As already indicated, the introduction of marketed output encouraged an increase in the share of material costs in marketed output. As could be expected in an administrative system, the authorities reacted to this abuse by introducing a new indicator. Since 1972 industrial and construction enterprises have been assigned a so-called guideline for the allowable share of material costs in one crown of marketed output. If enterprises exceed the planned share of material costs, the wage-bill is accordingly reduced.[37]

Optimistic expectations that the new provision would bring about a significant change in the behaviour of enterprises did materialise, but only partially. It soon became clear that the loophole was not fully closed, and therefore in 1976 the government resorted to new provisions. Industrial enterprises, where the indicator also includes the value of unfinished production and inventories of goods, are subject to a planned maximum of both. Similarly, construction enterprises are assigned a maximum target for the amount of unfinished work. A failure to observe these targets entails a reduction in the allotted funds for wages.[38] It can be assumed that even the new provisions will not bring the desired solution to the problem. Enterprises will certainly find new loopholes. Supervisory bodies have not and cannot have sufficient statistical data and other information to be able to make objective decisions about the allowable maximum share of unfinished products and inventories in the total volume of output. In addition the whole procedure is time-consuming and costly.

The performance of enterprises has also been recently evaluated according to the fulfilment of the planned relationship between growth of wages and productivity.[39] Though this is not a binding indicator, enterprises are under strong pressure to fulfil it.

The difficulties with wage regulation were one of the main reasons why Czechoslovak leaders decided to consider modifications of the system of management. On the basis of a decision by the government in December 1977, a small number of enterprises is now engaged in an experiment termed 'Complex experiment in management of efficiency and quality'.[40] The experiment includes planning, management, financing of investment, regulations of wages, etc.; and its main purpose is—as its name already indicates—to find ways to promote efficiency and quality. This is to be achieved by improving planning, putting greater stress on qualitative indicators and by extending the time horizon for decision-making.[41]

The branch ministries assign to experimenting enterprises limits for wages for the period 1978-80 and for individual years, divided into two components: funds for basic wages (basic wage-bill) and for bonuses. The first component is determined as before, with the difference that marketed output as an indicator has been replaced by net output.[42, 43] This change has been advocated for some time by Czechoslovak economists,[44] but apparently the planners preferred to wait for some Soviet initiative. The second component, which replaces the bonus fund to a great degree, is set as an absolute sum. Its use is conditioned by the fulfilment of planned productivity of fixed assets and of two other planned indicators (such as deliveries of export goods to capitalist countries, deliveries to the domestic market, etc.).[45] Overfulfilment of the plan is rewarded with a smaller adjustment coefficient than under-fulfilment.

The 'fond odměn' which was previously the bonus fund (and still is for non-experimenting units) has not been abolished. It seems that it will turn into a fund from which primarily year-end rewards will be paid, whereas the incentive component will encompass all other bonuses, including those which before were paid from the basic wage-bill. Whereas the incentive component is included in production costs, the year-end rewards fund will be fed from profit and other sources.[46, 47]

For an evaluation of the modifications, more information about the different aspects of wage and bonus regulation is needed. What is already clear is that the modifications do not present any radical change; they do not go beyond the present systematic frame.

As in some other countries the National Bank monitors the disbursement of wages by enterprises. If the local branch discovers that an enterprise is exceeding the set wage limit during the financial year, it is supposed to inform the enterprise's supervisory body of this fact. The supervisory body has to impose corrective measures on the enterprise; it can order a revision of the work norms, ask the Bank for a monthly check on the enterprise's disbursement of wages, or limit the enterprise's right to dispose of the funds in its bank account. If an enterprise overdraws its annual wage-bill and the overdraft cannot be covered from the reserve funds of the supervisory body, it is obliged to pay a charge to the budget from the bonus fund to the extent of the overexpenditure. In the event of an enterprise not having sufficient funds for payment of wages, the Bank will extend a repayable loan on which interest must be paid.[48]

Part Four

11 The Effectiveness of the Wage Regulation Systems as Anti-Inflationary Tools

Up to now we have devoted much of our attention to the so-called stimulative function of wage regulation. This has not been accidental; most of the countries under review experience greater difficulties with the consequences of the stimulative function than they do with the regulative function. For example Hungary carried out several modifications in its SWR, not because it was not able to cope with wage inflation, but primarily for stimulative and wage-differentiation reasons. To make the SWR effective it is necessary to strike a reasonable balance between its regulative and stimulative functions. It is not difficult to design a system which would limit wage growth to the plan targets; the crux of the problem is to design a system where the regulative and stimulative functions do not conflict, but instead further each other.

In practice the two functions need not act in the same direction. The objectives of the plan may not necessarily allow the pace for wage increases which is required for achieving stimulative purposes. This may be a result of objective conditions, but it may also be a consequence of the planners' wrong assessment of the pace of wage growth needed for accomplishing a certain stimulative effect. Also the methods used for wage regulation may conflict with the stimulation sought for. On the other hand, the built-in stimulative function may make it more difficult to keep wages within the limits of the plan—as will be shown later.

In this chapter we will attempt to analyse the effectiveness of the regulative function of the existing SWRs in the countries under review, in other words, the effectiveness of SWRs as anti-inflationary tools. Needless to say, since both functions of the SWRs are interconnected, this analysis must necessarily also deal with the consequences which the stimulative function has for the effort to fight wage inflation. Examination of the SWRs was confined to enterprises in the *khozraschet* sphere, mainly in industry. Therefore discussion of the effectiveness of SWRs will be limited to the same sphere.

First, it is necessary to explain which phenomenon can be regarded as inflationary at the enterprise level. It is obvious that in this regard the given definition of inflation in Chapter 2 must also be relevant for enterprises. There wage inflation was defined as excessive wage increases in relation to the rate of increase in the volume of goods and services. With some simplification it can be argued that enterprises contribute to inflation if their planned wage-bill is overdrawn. An important qualification should be made to this statement. As already stated, not every over-expenditure can be automatically classified as inflationary. It would be correct to do so if all other planned targets relevant to market equilibrium were merely fulfilled. In other words over-expenditure must be examined in the light of the overall fulfilment of the enterprises' plans. Therefore, over-expenditure may not be inflationary if it results from an over-fulfilment of plan targets relevant for market equilibrium.

As is known, over-expenditure of the planned wage-bill may be a result of a faster than planned growth of employment or of average wages or a combination of both. Faster growth of employment or of average wages may result from factors which are intrinsic to the SWR (due to a contradiction between its stimulative and regulative functions) or may result from factors which are outside the SWR. We will discuss both groups, putting greater stress on the first.

As has been shown, the five countries under review have different SWRs, and these are subsystems of the appropriate systems of management of the economy. This means, on the one hand, that our task cannot be accomplished by examining the effectiveness of the SWR generally; instead, a comparative evaluation of the systems must be given. On the other hand, there is a need to examine SWRs in their close interaction with the systems of management. In other words, the examination of SWRs cannot be confined to the systems themselves, but must also touch on the environment within which the SWRs must work. Therefore, we will first discuss some aspects of the systems of management which have an important bearing on the SWRs as anti-inflationary tools, and only then will we try to examine the SWRs themselves.

SYSTEMS OF MANAGEMENT AND INFLATION

To begin with, which of the two systems of management (we assume here only centralised and decentralised systems) is, due to its underlying philosophy and its way of functioning, generally better equipped to cope with inflationary pressures, or, more precisely, to successfully combat excessive wage growth? At first glance, it seems that a decentralised system is more vulnerable to upward wage pressures than an administrative system. A country with such a system cannot simply resort to orders and prohibitions without undermining the sensitive fabric of the decentralised mechanism. Neither can it very often change the rules of the game in order to cope with new situations. Frequent changes not only reduce the credibility of the government and hinder the process of stabilisation of the decentralised system, but they also introduce elements of uncertainty which make forecasting the behaviour of enterprises more difficult.

One of the characteristic features of a decentralised system is that it relies to a much greater extent than an administrative system on managerial initiative and incentives. Whereas planners in an administrative system are obsessed with the idea that only instructions, orders and strict control can make the system workable, planners in a decentralised system see in these methods a supplement to be used if incentives fail to induce the behaviour desired. Hence, a decentralised system must let wages and bonuses play a more stimulative role, which is only possible if there are no great strains on wage growth. A relatively faster wage growth is also necessary to allow a greater differentiation of wages, the need for which is felt much more in a decentralised system which expects improvement in efficiency to be primarily the result of the initiative of individuals. However, a widening of wage differentials at a period of small wage increases would mean no increase or even a decrease of wages for those whose relative position is to be worsened. And, as the Hungarian government experienced in the first years of the reform, this may cause political trouble.

It has already been mentioned that trade unions play a different role in the East than in the West. Yet there is a difference between the role of TUs in a centralised and a decentralised system. A decentralised system is more willing and more under pressure to let the interests of individual groups come into the open; therefore it grants TUs more scope for their traditional function. It is know that the relatively high wage increases in Czechoslovakia in 1968–9 were also due to the pressure of the TUs. The Hungarian government's interference with wages in 1973 which brought about higher wages, primarily for blue-collar workers, was initiated by the TUs.[1]

In a decentralised system where elements of the market mechanism are integrated in the system, it is difficult to keep price movements within wanted limits, at least, more difficult than in a centralised system where prices are fixed by the centre. A decentralised system cannot do without some price flexibility, that is without decentralising price formation to a certain degree. And this is, as the Hungarian and Czechoslovak examples show, inflationary.[2] 'All decentralization, then,'—writes P. Wiles—'is inflationary; central price control is more likely to work. No wonder that Communist ministers of finance who are responsible in this field have always resisted reforms'.[3] In addition, as was shown in Chapter 1, countries with a decentralised system are more inclined to use price increases to solve problems of price distortions and to equalise incomes of earners in agriculture with those in non-agricultural sectors. Naturally, price increases prompt demands for wage increases.

This is not to say that administrative systems have no problems in coping with price pressures. Reference has already been made to the fact that enterprises try to circumvent price fixing by the centre by various methods, one being under cover of the introduction of new products. Higher prices are demanded on the pretext that improvements have been made, though these improvements are in reality less costly than the difference between the new and old price tags. Price agencies are not in a position to check on all the cost calculations which enterprises are obliged to submit with their demand for a higher price; therefore, they necessarily rely heavily on the enterprises' own estimates.

Up to this point we have advanced arguments which could be briefly summarized by saying—as P. Wiles does—that decentralisation is inflationary.[4] The picture will not be accurate if we do not mention other systemic differences which may influence the behaviour of enterprises. It can be generally argued, and the examination of the SWRs has shown and will show it more clearly, that a decentralised system is more geared to consumer demand and more conditioned for improvement in economic efficiency than a centralised one. It is known that in the centralised system many provisions, orders and instructions which are designed to enhance attention to consumer demand and to improve efficiency are frustrated. Despite strict controls enterprises often find loopholes which allow them to circumvent provisions which contravene their interests. The reluctance of central planners in an administrative system to take adequate consideration of the interests of managers and other personnel of enterprises is the main reason for this.

THE EFFECTIVENESS OF SWRs

Turning our attention to the effectiveness of the three SWRs, we must first make clear that there is no way to quantify the effectiveness of wage controls in the fashion practiced in the West where the effectiveness of controls is usually measured by how much inflation rates were reduced through control. The actual rate of inflation is compared with the hypothetical rate which would have arisen if not for the controls. Some economists also try to figure out the costs of the controls to the economy in order to find out whether such controls are worthwhile. In the Soviet Union such computations are not feasible, simply because controls have been in effect there for four decades; and attempts to compare, let us say, the existing rates of open and repressed inflation with a potential situation without control would be an exercise in futility. The same is true with regard to East European countries, even if controls there have been in operation for a shorter period of time.

In our opinion the best way to tackle this problem would be to compare the actual growth of wages in industry or several branches of industry with the planned wage targets. But even here caution is warranted. Reference has already been made to the fact that not every over-expenditure can be classified as inflationary. To obtain a really reliable picture of inflationary pressures in industry or several branches of industry would require a thorough analysis of the performance of enterprises, which can be done only by insiders who have access to all the needed statistical data. It is not feasible to implement even a less ambitious goal—such as a simple comparison of the actual growth of wages in industry with plan targets—because of a lack of adequate figures for all the five countries involved.[5] Therefore we will try to tackle this problem indirectly by proxy, by analysing the SWRs from the viewpoint of two criteria:

1. To what extent the individual SWRs orient enterprises to a satisfaction of consumer demand. It can be assumed that the less a system is geared to consumer demand, the greater is its destabilising effect on the market for consumer goods and the more it is prone to inflationary pressures.

2. To what extent the built-in mechanism in individual systems gives protection against tendencies to over-expend the planned wage-bill in enterprises.

The second criterion is certainly the more important one because it gives a direct answer to the problem raised. However, to ignore the first criterion would mean neglecting an important aspect of the stimulative function of SWRs which has its bearing on the regulative function.

SWRs and Consumer Demand

Let us start with the direct system. It seems to have been sufficiently demonstrated that it orients enterprises to a fulfilment of assigned output targets rather than to a satisfaction of consumer demand. This is the direct result of linking wage-bill growth to the fulfilment of a success indicator. This is not to say that this linkage forces enterprises to produce products in a certain detailed mix which, due to the planners' known inability to predict the evolution of demand precisely, necessarily results in imbalances between supply and demand. It is generally known that central planners, for employment and technical reasons, are not in a position, even if they wished to do so, to hand down a detailed product mix to all enterprises. Therefore in most enterprises managers have some room for manoeuvre, and within this scope they could do a lot for consumers. However, the crux of the problem is that the system is designed in a way that makes the managers concerned primarily with achieving the success indicator with the smallest effort possible, because this is the way to maximise the wage-bill. In other words, growth of the wage-bill is not directly linked to the satisfaction of consumer demand but to the degree of fulfilment of the plan targets expressed in an indicator. Satisfaction of consumer demand is a secondary task. No doubt, if the pursuit of wage-bill maximisation coincides with consumer demand, there is no reason for managers to act against consumers' interests. However, if they are in conflict, managers will not hesitate to let the interest of the enterprises prevail over the consumers'.

Gross value of output (and to a great degree marketed output as well, as used in Czechoslovakia)[6] strengthens this tendency. As the term itself indicates, it is an output indicator and therefore makes enterprises focus on output. Sales are primarily the concern of government agencies. This arrangement is not accidental. It is an integral part of the concept of the administrative management system, which is based on the belief that enterprises are more motivated to fulfil the assigned output targets if they are not concerned with sale of goods produced by them.

We are aware that the statement about gross value of output can be challenged by arguing that under conditions of a seller's market it makes no difference which indicator is used. In other words, it will always be possible for enterprises to impose on their buyers the goods they produce and therefore avoid gearing their output to demand if this does not serve their interest. In our opinion, a genuine sales indicator could partially alter the situation even in conditions of a seller's market. Enterprises in their capacity of suppliers would be forced to give more thought to the wishes of the consumers. It is, however, questionable whether a centralised system can afford to allow a genuine sales indicator. In brief, the linkage of the wage-bill to a planned indicator, particularly if this indicator is gross value of output, tends to have a destabilising impact on market equilibrium. This tendency is strengthened by the inefficiency which this linkage produces. As has already been mentioned, it makes enterprises reluctant to mobilise all their resources for a maximum effort (see p. 116).

The picture would not be complete if mention was not made of the fact that direct incentive systems have built-in devices to counter the adverse effects of linking the wage-bill to gross value of output. As a reminder, we will mention some of the devices again. In Czechoslovakia and the GDR, and to a lesser degree in the USSR, the size of the bonus fund is also dependent on the amount of produced profit, a linkage which is supposed to prompt enterprises to become more efficient and also to have more regard for consumer demand. In the Soviet incentive system, there are special allocations to the bonus fund for overfulfilment of plan targets in productivity and for the share of high quality goods in the total volume of output. Similar incentives exist in the Czechoslovak and GDR systems. The attempts to apply long term normatives in the USSR are also supposed to act in the direction mentioned. There is no way to measure the offsetting effects of the incentive systems. What is clear is that they cannot change essentially[7] the adverse effects of linking the wage-bill to plan targets which are expressed in terms of gross value of output or similar indicators.

We turn now to the systems in which wage growth is linked to an indicator which measures performance in terms of a comparison with the previous year. This is a shared feature of what we called the mixed system (the Polish) and the Hungarian system which we characterised as close to indirect. A question may be raised: to what extent are these systems geared to consumer demand? There is a difference between the Hungarian and Polish systems. In the Hungarian system binding targets, with some exceptions, have been dismantled.[8] The performance of enterprises—as reflected in the evaluation indicators—depends in principle on the degree to which they are capable of satisfying demand. For this purpose the regulators of wage growth and, to a greater extent, of the bonus fund are designed in a way to gear enterprises to market demand. Both main indicators of wage growth—gross income per employee and value added—are net and sales indicators at the same time. Needless to say, produced profit, which is relevant for the size of the bonus fund, is a sales indicator. In addition to this, there is in the Hungarian system a dual linkage between wage increases and profit. On the one hand, the regulators of the growth of wages themselves depend on profit (which is a part of gross income and value added) and, on the other hand, the ability to pay taxes for wage increases depends on the size of produced profit. (As was already mentioned, taxes to be paid for wage increases come from the bonus fund.) And profit in turn depends on the size of sales among other things. Therefore it is fair to assert that the performance of enterprises in the Hungarian system is measured by yardsticks which reflect the degree to which enterprises are capable of satisfying demand.

The statements made are confirmed by news reports and reports of tourists who have visited Hungary or Hungarian tourists who have travelled abroad. No doubt Hungary, among all the Soviet bloc countries, has made the greatest progress in curing the old disease characteristic of the East: permanent consumer goods shortages and low quality of goods. This is not to say that it has managed to solve all the problems in this area; for example it is still plagued by a noticeable seller's market.[9]

In Poland the 1973 reform retained binding targets but not to the pre-reform extent. In addition the central planners committed themselves to reduce gradually the number of binding targets and to replace them by economic instruments (parameters and normatives).[10] What is also of importance is that the regulator of wage growth is—as in the Hungarian system—a net and a sales indicator at the same time. However, the role of profit in the regulation of wages is smaller than in the Hungarian system; it is confined to being one of the elements of output added. Nevertheless, the SWR orients enterprises more to consumer demand than a traditional centralised system. It should also be kept in mind that the size of the managerial bonus fund depends on profit. Market disequilibrium problems, which Poland has been experiencing recently, are due to other causes, one of which is the inability of the Polish authorities to tackle the disparities between prices of food and other consumer goods (see Chapter 1). Another cause, which has its origin in the SWR itself, will be mentioned below.

SWRs and Planned Wage-Bills

Turning to the second criterion, one should bear in mind that managers as well as workers—regardless of systems—are interested in maximising wages and bonuses.[11] 'Experience shows,' writes B. Fick, 'that enterprises consistently aim at maximisation of wages by all possible means. There is nothing in their internal structure that would resist such pressures'.[12] He mentions several reasons for this with which it is possible to agree. The fact that an enterprise can pay higher wages and bonuses than others enables that enterprise to attract more qualified personnel, reduce labour turnover and create better relations between management and workers. Enterprises also strive for more funds for wages because this facilitates keeping the labour reserves needed in an administrative system for an easy fulfilment of plans.[13] There is a tendency in enterprises not only to maximise average employment incomes but also to maximise the wage-bill, which need not necessarily result in higher employment incomes. Enterprises have a tendency to hoard labour, and this can be achieved only if they succeed in increasing the wage-bill.

On the basis of the foregoing statements it is legitimate to ask the questions: what are the means and ways by which enterprises try to achieve their mentioned objective and how well do individual systems cope with it? To begin with—the tendency to hoard labour is an important factor in the upward pressure on wage-bills. Most of the reasons for this phenomenon, which is not limited to the direct system, were discussed in Chapter 2. Here another is added, which is stronger in a centralised system and which has a twofold effect: it not only increases the demand for labour but is also a factor in pushing up average wages. This is the well-known feature: the uneven distribution of the workload in enterprises during the month and year, and the effort at the end of each period to catch up with plan targets. This so-called *shturmovshchina*, which is primarily a result of an uneven supply of materials during the year and taut plans, leads to overtime work including 'special brigades' on Sundays and other holidays and to wage outlays higher than envisaged in the plan. This may be a source of market disequilibrium, all the more since heavy industry prevails in all the countries, and growing wage costs, even if they were covered by increased production, are not offset by a corresponding expansion of the supply of consumer goods and services.[14]

All the countries of the Soviet bloc are trying to cope with labour hoarding. The methods used differ: in the USSR, the GDR and Czechoslovakia, the planners are endeavouring to achieve this goal by stimulating increases in labour productivity. In addition, in Poland employment increases are subject to payments into the branch reserve fund. And finally, Hungary is trying to tackle the problem among others by extending wage-bill regulation.

Let us turn to the discussion of the pressures for wage-bill maximisation from sources other than employment. The widespread piece rate system is one of the significant factors. Most of the work norms—as already referred to—are in fact set in enterprises and this gives piece rate workers the opportunity of exerting pressure on the norm-setting staff for a slackening of work norms. The more sophisticated and unique the job performed, the greater is the chance of the workers to have their demands met at least halfway. The period of '*shturmovshchina*' is the most suitable time for workers for pushing through advantageous work norms. There is also another circumstance which allows the piece rate system to be used as an instrument for pushing wages upward, namely, the principle adopted in the countries with an administrative system that wage rates should not change frequently. As already explained in Chapter 3, this leads to a slackening of norms, a process which offers workers a good opportunity to press for higher wages.

Another reason why the piece rate system contains the seeds of inflation is that in the vast majority of cases the so-called straight piece rates are used, which means that the reward depends on the set time norm (for the performance of a certain operation) times the wage rate.[15] In other words, the reward is proportional to the degree of fulfilment of the work norms.[16] If the over-fulfilment is high, this may push up average wages beyond the plan targets. In addition, the straight piece-rates are an incentive to over-fulfilment output targets, which will be discussed later on. Recently, there has been an effort to cope with the unfavourable consequences of straight piece-rates by introducing other wage forms[17] and also by experimenting with regressive piece-rates.[18] The widespread piece-rate system is one of the main reasons for the high share of low quality goods, including rejects, in the total volume of output, a disease which the Soviet Union in particular is wrestling with.[19] A piece-rate system focuses workers' attention primarily on quantity at the expense of quality and encourages an excessive use of materials and investments.

One way to achieve a higher wage-bill as well as higher average wages is through over-fulfilling output targets. Over-fulfilment of targets is an attractive goal to strive for; it is still rewarded more favourably than fulfilment of targets. This is true even now when the adjustment coefficient for over-fulfilment is no longer unity as it was in the past (see Chapter 7). There are several reasons why such an over-fulfilment, which is encouraged by the direct system where the actual size of the wage-bill depends on the degree of the fulfilment of output targets, may unfavourably affect market equilibrium. Not only may an increase in the wage-bill boost employment unnecessarily or else not be matched with adequate output, but also it may result in the production of unsaleable goods. In other words, additional purchasing power may be created without being matched with an adequate supply of goods and thus inflation (open or repressed) may be generated.

All three countries which have this system have recently been trying to use new methods to cope with its consequences. In the USSR the State Bank is empowered, as already stated, to reduce the adjustment coefficient by 50 per cent if the over-fulfilment is achieved by not observing the targets in productivity. It is obvious that this does not solve the problem; it may induce enterprises to increase productivity, but it does not prevent them from producing unsaleable goods. The Czechoslovak provision of at least two adjustment coefficients for over-fulfilment of targets (one, higher, for over-fulfilment up to a certain limit and the other, lower, for over-fulfilment above this limit) has the same shortcoming. The GDR is trying to tackle this problem by a method similar to the one applied by the Polish planners in the sixties. Only over-fulfilment of targets which is deemed necessary from a national viewpoint is rewarded.

Up to this point we have discussed mostly the direct system. Now we must turn to the non-direct systems. It can be assumed that there the piece-rate system exerts upward pressure on wages as in the direct system. Little concrete information is available on this topic. What is known is that in the Hungarian system the progressive taxation on wage increases has forced enterprises to reduce the role of the piece-rate system and, where it has been retained, to extend regressive piece-rates.[20]

The author feels that he is on firmer ground when the question of the impact of the wage rate system on norms is raised. In Hungary and Poland the general practice in countries with an administrative system—to let wage rates remain rigid for a certain period of time—has been replaced with a more flexible solution (see Chapter 3). Therefore it is safe to argue that for this reason neither of them faces pressure for slackening of work norms (and its consequences for the growth of wages) to the degree that the USSR and Czechoslovakia do.

One way to maximise wages with the smallest effort is to juggle the indicator which regulates wage growth. No system is safe from such activities. There are in substance three groups of methods of juggling indicators. One is to include in the value of the indicator activities which are not the enterprise's own work. The second one consists of shifts in the output mix to products which increase, without enhanced labour input, the final value of the indicator. And, finally, the third method lies in juggling prices.

The first method can be applied only in the direct system which is marked by gross indicators, whereas the other two are used in all systems. There is a universal tendency to shifts to more profitable products and, in this way, to increase the amount of produced profit. In non-direct systems where net indicators are used, such a shift has much greater influence on the wage growth regulator than in a direct system with a gross indicator. In the latter, it is more beneficial to take advantage of shifts in output mix to more material intensive products. Manipulating prices is also practiced in all the systems.

The possibility of juggling indicators is of concern to planners in all countries, particularly in Poland. As already indicated in Chapter 8, it turned out that the failure to make output added an objective yardstick is the weakest point of the system. The Soviets do not really have an effective tool for discouraging enterprises from undesirable shifts in product mix. The Czechoslovak planners try to avoid shifts to more material intensive products by imposing on enterprises a normative for the share of material costs in marketed output. Since these efforts have not turned out to be effective, Czechoslovakia, following the Soviet pattern, is experimenting with net output. The Hungarian and Polish planners are in a better position in this regard; they rely on the wage-bill regulator being a sales indicator to act as a barrier to great shifts to more profitable products motivated by the desire to maximise wages.

Some countries (Hungary and Poland, as well as the USSR) try to cope with price juggling by taxing away or confiscating (USSR)[21] the profit produced in such an unjustified way. In addition Hungarian and Polish wage regulation systems have built-in instruments which aim at reducing the impact of such activities. In Hungary it is the progressive taxation which puts limits to the translating of price juggling into wage increases, whereas in Poland the same role is attributed to the recently introduced contributions to the branch reserve funds.

The Hungarian and Polish systems offer some protection against the well-known disease of the centralised system—to produce unsaleable goods. Wage growth is not linked to the fulfilment of plan targets as in the direct system, and there is no special incentive to exceed a certain level of performance (regardless of the salability of produced goods). It should be borne in mind that in both countries the regulator of wage growth is a sales indicator, and this fact surely acts as a discouragement to the production of unsaleable goods.

The examination up to now has confirmed Fick's assertion that enterprises try to maximise wages by all means and do not hesitate to take advantage of any loophole in the system. But it has also shown that the vulnerability of the systems to the different means used varies, depending on their structural coherence, i.e. the degree of consistency of the stimulative and regulative functions. Generally it can be said that the direct system is structurally less coherent and more torn by contradictions. Enterprise's desire to over-fulfil targets even at the cost of producing only for the warehouse (and, thus, contributing to a generation of market disequilibrium) results, undoubtedly, from inconsistencies between the stimulative and regulative functions.

At this point we can turn our attention to a discussion of the gist of the second criterion. The direct system has the 'advantage' that only little is left to chance. The greatest part of the basic wage-bill of an enterprise (and the same is true of the bonus fund) is fixed from the centre. Judging on the basis of the fulfilment of industrial output plans, we can assume that the difference between the size of the planned wage-bill for industry as a whole and the actual size authorised is small. Of course in some enterprises this difference may be large. The direct SWR, including Bank control and other administrative controls, has proved quite effective though not watertight. From the figures indicated in Tables 2.5 and 2.6 it can be concluded that the planned wage fund on the national scale was often overdrawn. Many complaints about wage-bill over-expenditures in the Soviet literature as well as in the Czechoslovak are further evidence that the direct SWR is far from perfect. It would, however, be incorrect to regard every over-expenditure as inflationary for reasons already mentioned. What is of no less importance is that even inflationary over-expenditures are small (negligible by Western standards) and by no means of an extent to destabilise the economy seriously. Furthermore, much of the over-expenditure in the USSR and to a lesser degree in Czechoslovakia results from faster growth of employment than planned for (see Table 2.6).

The repressed inflation existing in the USSR, the GDR and Czechoslovakia is primarily the result of factors other than wage inflation (particularly if we mean by this excessive growth of total average wages). To the extent that wages contribute to repressed inflation, they do so because the SWR in its stimulative function (mainly with regard to over-fulfilment of output targets) gives rise to imbalances between supply and demand for certain groups of consumer goods.

In the non-direct systems the wage-bill is not fixed from the centre. Nevertheless the Hungarian system works well. This is true to a lesser extent of the Polish system. In the Hungarian system the normative is low on the average. P. Marton asserts that in practice an enterprise with wage-bill regulation and an unchanged number of employees which wants to increase wages by 5 per cent must achieve a 12 per cent increase in profit, which is one of the two important parts of value added; in the case of a $5\frac{1}{2}$ per cent increase in wages, the corresponding increase in profit must be as high as 19 per cent.[22]

But this is only the first line of defence against excessive wage increases. The second defence line lies in taxes which are characteristic of an indirect system. True, we characterised an indirect system as a system where taxation is used as the only wage regulator. Such a system existed in Czechoslovakia for a short time, but did not prove itself. Here it should be added that the main reason for its failure was the application of a single tax rate. In January 1969 the government replaced the single tax rate by a progressive one, but due to Dubček's ousting in April 1969, there was really no time to test the new measure.

As already mentioned, the Hungarian planners have proved to be more pragmatic. They do not rely only on taxes but use them as a second defence line, the first being the linkage of wage growth to economic performance. What is no less important is that the taxes are mostly heavily progressive. The development of wages in the last two years is perhaps the best proof that the Hungarians have found in this combination, in which taxes play a very significant role, an effective tool for controlling wages. Against different odds, taxation has proved to be an important barrier which has prevented wages from increasing much above the expected limits. Despite faster increases in profit and accumulation of reserve funds in enterprises than anticipated, and a faster increase in prices than envisaged in the plan, taxes have managed to resist the tide of wage increases.

In the first half of the seventies the planners had difficulty striking a correct balance between making the tax a barrier against unreasonable wage increases and a burden which made it impossible for enterprises to give their personnel a reasonable wage raise. They were too cautious, and the heavy taxation which they imposed involved many enterprises in great difficulties. Interministerial committees were flooded with applications for tax exemptions. Heavy taxation was one of the reasons why the economic autonomy which the reform was supposed to promote could not materialise to the expected degree.[23] The modification in taxation in 1976 is no doubt a step in the right direction. Taxes are no longer paid for every increase in wages, but are limited to wage increases above 6 per cent and above the amount to which enterprises are entitled on the basis of their performance (including guaranteed allowance). The differentiation of taxation, the application of a proportional tax on wage increases above 6 per cent which are deserved by performance, and the imposition of a progressive tax on wage increases above performance limits aim at being a stimulus to higher effort. With these modifications the tax has become a real second defence line also in the sense that it fulfills a different function than the first line. The latter fulfills primarily the stimulative function and the former primarily the regulative one. This division of functions essentially reduces the possibility of contradictions between the two functions.

In Poland the normatives were set at a higher level than in Hungary, the purpose of which was apparently to enable enterprises to accelerate the rate of wage growth. It turned out that enterprises managed not only to ensure unprecedented high wage increases (the biggest since 1956) but also to accumulate considerable wage reserves. (Apparently many enterprises managed to increase output added at a much higher rate than was anticipated. It is not known to what degree these results were influenced by juggling, but since the authorities introduced measures to cope with such a danger, it can be assumed that it had a share in the results.) Confronted with this situation, the authorities introduced a supplementary wage control in the form of contributions to the branch reserve fund. As already mentioned, their progressivity is mild compared to Hungarian taxes, and it is questionable whether they will be sufficient to cure the situation. Interestingly enough, the Poles have not yet tried to use taxation as a regulator of wages to any great extent though some economists have called for it.[24] Perhaps the fact that the Polish planners always resort to the use of reserve funds when something goes wrong with wages makes a tradition which it is difficult to abandon.

In summary, it can be said that all the wage control systems work more or less well, but yet the Hungarian system seems to be superior to the direct system. If we consider the comparison simply from the viewpoint of how individual systems cope with wage inflation, then it is not an easy task to substantiate this statement. One could question it, arguing that the direct system (whether the USSR, the GDR or Czechoslovakia is considered) has stood the test well. Yet it should not be forgotten that anti-inflationary policy in all those countries is also exercised through a rigid price policy. On the other hand, the Hungarian SWR has to work in an environment of a flexible price policy. It should, however, be made clear that upwards creeping prices in Hungary are the result of price policy and other factors and not of the ineffectiveness of wage control (see Chapter 1). The Hungarian linkage of wage growth to performance (we disregard the direct controls applied in some sectors of the economy) supplemented by heavy taxation has proved to be an effective system. In our view, what is undoubtedly superior in the Hungarian system is the fact that the SWR is marked by greater structural consistency. In contrast to the direct system, the Hungarian in its stimulative function is more geared to satisfying demand and promoting efficiency, and therefore more effective in helping maintain market equilibrium.

Conclusion

In the conclusion to this study we will briefly attempt to sketch the trends in wage regulation as they appear to us and pinpoint problems which still wait for a solution in order to make the SWR more effective in all its functions. It is obvious that the present systems cannot be regarded as definitive. It has been shown that the non-direct systems are in a state of flux; even the Hungarian system, in our opinion the most effective in the Soviet bloc, is gradually undergoing changes. Moreover the recent debates on the effectiveness of the system in Hungary show that even suggestions are being made for a radical overhaul of the system.[1] The Soviet SWR too, which up to now has displayed a special immunity to pressures for change, now seems bent on some modifications.

Looking at the directions in which these systems are moving, one cannot escape the impression that a tendency to convergency manifests itself. It appears that the Soviets will ultimately replace gross indicators with net indicators. However, it is not likely that the new net indicators will be designed as genuine sales indicators. Judging from the current trend in the USSR toward increased centralisation, it would be unrealistic to expect any fundamental changes in the SWR such as the abandonment of the present linkage of wage growth to the fulfilment of plan targets. It can be assumed that if the Soviets decide to opt for net indicators, Czechoslovakia (which has already embarked on an experiment with them) and the GDR will follow suit. On the other hand, the Hungarian planners will probably continue in their effort to adjust the SWR more to the conditions of individual sectors.

Differentiation will thus gradually gain more ground at the expense of uniformity. It is not inconceivable that the Hungarians may expand direct regulation of wages at the expense of 'indirect'. Changes in Poland can also be looked for; the system does not work as well as expected. It appears that the changes will also be in the direction of increased centralisation.

It will not be surprising if the tendency to convergency will assert itself in another area also: the use of taxation as a tool for regulation of wages. The quite successful utilisation of this tool by Hungarian planners may encourage other countries to follow suit. The recent introduction of charges on wage-bill increases in Poland is close to the idea of taxation. Needless to say, taxation cannot play the same role in an administrative system as it does in Hungary. It is questionable whether, for example, the USSR is prepared to give enterprises the economic and financial autonomy needed to make taxation an effective tool of wage regulation. But even so, taxation may help to strengthen the regulative function of the direct system without harming the stimulative function.

Turning to the second point, we think that the most important problem which all the countries face in the field of wage regulation is how to make the success indicator more objective. It is obvious that the adoption of a net indicator does not solve the problem, though it creates better conditions for its solution. Disregarding the various ways of juggling with indicators, the objectivity of an indicator is conditioned by the degree of rationality of the price system. All the countries under review, to a greater or lesser degree, suffer from price distortions. Even the Hungarians—despite great efforts—have not managed to bring into being a rational price system. The need for a sweeping reform of the price system (wholesale and retail) is felt in all the countries, mainly in Poland and Hungary, for reasons already explained. However, due to the possible political repercussions, it can be assumed that no radical changes will occur, especially not in the near future.

Another important problem which confronts Hungary particularly is how to stimulate greater effort by wage and bonus regulation and, at the same time, avoid a widening of wage differentials which may be politically unacceptable. In the Hungarian system more than in the Polish (not to mention the direct system), profit still plays a very important role. This is the reason why some economists are calling for a break in the linkage between profit and wages. Should the planners accept such a suggestion, it would mean a greater application of direct methods in the determination of wage growth. Such a course would necessarily affect the economic autonomy of enterprises adversely.

Another no less burning problem is how to use the SWR for a better utilisation of labour.[2] At issue is not only how to stimulate enterprises to higher increases in productivity with the existing work force, but even more how to achieve a more balanced allocation of labour among enterprises. It is known that in all the countries many enterprises suffer from labour shortages when, at the same time, other enterprises do not fully utilise their work force. All the countries try to tackle the problem, but for the time being no effective solution has been found.

In particular, the countries with a direct system still have to wrestle with the problem of how to induce enterprises to an expansion of output without generating inflationary pressures. Here several problems which have already been mentioned are involved. To reiterate, specifically: how to induce enterprises to reveal their reserves and how to transform their interest in over-fulfilment of plan targets into production of goods which are in demand. The methods used hitherto have not proven effective.

Notes

NOTES TO INTRODUCTION

1. In this study the terms control and regulation are used interchangeably though a case can be made that they differ in their meaning. Control has rather the connotation of a negative action (in this case to prevent excessive wage increases) whereas regulation has a positive connotation in the sense that it is also meant to be an incentive.
2. A. Katsenelinboigen, 1975, p. 101; B. Schwarz, in *Struktur-und stabilitätspolitische Probleme in alternativen Wirtschaftssystemen*, 1974, p. 121, gives a similar definition.
3. See *Anti-Inflationary Policies: East-West*, CESES, *L'Est*, no. 6, 1974 (hereafter referred to as CESES).
4. G. Garvy, 1974, p. 84.
5. P. Wiles, 1974, pp. 231–2.
6. P. Wiles, 1977, p. 371.
7. Ibid., p. 372.
8. B. Hornok (*Pénzügyi Szemle*, nos. 3 and 4, 1976) contends that with a faster growth of investment than of national income, which is mostly the case, inflationary processes start to show up.
9. O. Šik, 1973, p. 101.
10. See Z. Fallenbuchl, 1973(a), March–April 1973.
11. B. Hornok (see 8 above) concludes that a lessening of economic disequilibrium can be achieved primarily by a reduction of the rate of accumulation and the stock of unfinished investment projects.
12. See M. Gamarnikov, Sept.–Oct. 1972 and also G. Schroeder, 1975.
13. See W. Brus, 1972, p. 149.
14. We disregard here the usual practice of preferentially supplying big cities and industrial centres with consumer goods.
15. This statement is a rearrangement of G. Garvy's contention (1974, p. 302), with the deliberate intention of stressing the role of insufficient supply.

16. See L. Berri, 1973, vol. 2, pp. 158–9.
17. See J. Chapman, 1963, pp. 110–12; G. Schroeder, in *Soviet Economic Statistics*, 1972, p. 292; *Rocznik Statystyczny pracy 1945–1968*, 1970, p. 402; *Statistická ročenka*, p. 131 and *Narodnoe khoziaistvo SSSR*, 1972, p. 789.
18. Some money payments are not included in either the basic wage-bill or in the bonus fund. It seems that there are some differences in detail as to what is and what is not included in individual countries. See G. Schroeder, in *Soviet Economic Statistics*, 1972, p. 292 and J. Jílek et al., 1976, p. 265; *Rocznik Statystyczny*, 1976, p. 104 and *Közgazdasági Kislexikon*, 1968, p. 54.
19. The inquisitive reader is advised to read J. Zielinski, ch. 5, 1973, where he will find a detailed analysis of indicators based primarily on Polish experience.
20. The Germans group instead indicators into material and financial (value) ones which is a self explanatory distinction.
21. *Planirovanie narodnogo khoziaistva SSSR*, 1967, p. 40.
22. J. Zielinski, 1973, p. 193.

NOTES TO CHAPTER 1

1. See G. Grossman, 1977, p. 133.
2. For example, in 1968 Czechoslovakia and Hungary still had 10,000 different tax rates. A great deal of the differentiation has no rational substantiation.
3. J. Adamíček, *Politická ekonomie*, no. 8, 1968.
4. E.g., East Germany increased wholesale prices by 12% in the period 1964–6, Czechoslovakia by 29% in 1967 and the USSR also had some increases in 1966–7, but none of them increased retail prices at the same time. See Sztyber, *Ekonomista*, no. 6, 1969.
5. For more about this topic see J. Adam, 1974, pp. 108–13 and T. Nagy, 1960, p. 60.
6. F. Holzman, no. 2, May 1960.
7. I raised this question with a very well-known economist. His answer was that to ascribe such intentions to East European planners would be to overestimate their sophistication.
8. 'Although each Five-Year Plan scheduled a more or less substantial deflation,' writes N. Jasny (1951, p. 13), 'no real effort was made to accomplish this. Indeed, only during the 1st Plan period was a serious attempt made even to hold to the existing price levels.'
9. This is not to say that prices increased spontaneously. They were regulated from the centre like wages.
10. E. Zaleski, 1971, pp. 89–91.
11. M. Dobb, 1966, p. 242.
12. N. Jasny, 1961, p. 81.
13. M. Dobb, 1966, p. 243.
14. According to D. R. Hodgman (1954, p. 112) labour productivity in Soviet large scale industry declined after 1929 (it was 92% in 1932). Official

figures show a huge increase, though still far below the target. (See E. Zaleski, 1971, p. 249).
15. See M. Dobb, 1966, p. 239, and also A. Baykov, 1947, p. 342.
16. N. Jasny, 1961, p. 67.
17. See A. Baykov, 1947, pp. 33–50.
18. See also *Einheit*, no. 10, 1949.
19. *Stručný hospodářský vývoj Československa do roku 1955*, 1969, p. 47.
20. *W.W.I.—Mitteilungen*, Cologne, no. 12, 1958.
21. For more, see J. Typolt, V. Janza and M. Popelka, 1959, p. 279, and also J. Adam, 1974, pp. 104–7.
22. See *Press Review*, Basel, 22 March 1951, and *The Times*, London, 3 December 1951.
23. T. I. Berend, 1974, pp. 42–3, and B. Nogaro, September 1948.
24. T. I. Berend, 1974, p. 84.
25. Č. Kožušník, 1964, p. 101.
26. M. Kucharski, 1972, pp. 310 and 320; see also J. Montias, 1964, p. 233.
27. *Wochenbericht*, 6 November 1953.
28. All the countries carried out a currency reform aimed at nullifying the consequence of the Second World War. Czechoslovakia did so in 1945, Hungary in 1946, Poland in 1945 and East Germany in 1948.
29. E. Ames (no. 3, June 1954) argues that Poland (in 1949 and 1953) and Hungary in 1951 abolished dual prices without resorting to currency conversions, since the currency surplus was not so high as to require new prices to be close to the commercial prices.
30. Usually nationalisation and collectivisation are mentioned in Soviet and East European literature as two measures applied by all the People's democracies for the sake of changing ownership relations. In our opinion, currency reforms also played an important role, though of course, a much smaller one than the measures mentioned.
31. See also M. Spulber, 1957, p. 128.
32. For more, see *Tägliche Rundschau* (containing the official decree), 16 December 1947; E. Pithe, *Finanzarchiv*, no. 3, 1949 and Ia. Kronrod, ch. 3, 1950.
33. For more, see M. Kucharski, 1972, pp. 320–1; *Pressedienst der polnischen Militarmission*, 1 February 1950 and *Wirtschaft*, 1950, no. 45.
34. For more, see Typolt, Janza, Popelka, 1959, pp. 283–9; Č. Kožušník, 1964, pp. 96–114 and *Stručný hospodářský vývoj* . . . , 1969, pp. 474–80.
35. *Gesetzblatt der DDR*, part I, no. 73, of 13 October 1957 and *Neues Deutschland*, 14 October 1957.
   As a matter of principle, savings deposits in East Germany were converted into new money without questions being asked about their origin.
36. M. Kucharski (1972, p. 321) indicates the following figures for the social origin of cash holders who submitted money for conversion: 20.8% of the submitted amount belonged to workers, 38.3% to the rural population and 40.3% to the private urban sector.
37. In East Germany people who wanted to move to the West usually did not keep their money in the banks for fear that a withdrawal of their money before departure might attract the attention of the police and thwart their plans, as happened in some cases.

38. E.g., in the Czechoslovak currency reform the price of rice was fixed, at Kčs28 for one kg (approx. $4) and was gradually reduced to Kčs5.
39. See A. Nove, 1972, p. 305.
40. Among ideological considerations, mention should be made of the prevailing notion, allegedly Marxist, that with growing productivity the drop in the value of goods should be reflected in price decreases. (See B. Csikós-Nagy, *Valóság*, VI, 1974.) The relatively widespread idea that the sources of the growth of productivity must determine the means of distributing its results, may have also contributed to the reliance on price cuts. It was believed that the gains in productivity which are a result of objective factors (such as investment and technological progress) should accrue to the whole society (and not just to employees in the factories where the gains were achieved) in the form of price reductions and tax revenues. It was so argued in the name of the principle of distribution according to labour.
41. *The Economist*, London, 10 April 1954. According to *Wirtschaft des Ostblocks*, no. 42, 1954, despite these price reductions, prices in 1954 were still approximately 100% above the level of 1940 whereas nominal wages were only 90% higher.
42. For more see M. Kucharski (1972, pp. 313–27) who gives a detailed analysis of the price evolution in the 1950s.
43. The figures indicated are official figures and should be treated with some reservations. It is known that in Czechoslovakia, for example, the size of savings gained by the population (which the government used to publish at each price reduction) was calculated on the basis of the volume of depreciated consumer goods put into circulation, without taking into consideration whether they were sold or not. Often goods were included in price reductions which enterprises intended to stop producing and the price reduction served as a means of clearing stocks, primarily of various industrial goods. Finally, there were often reverse cases where enterprises stopped producing cheapened goods and replaced them with others, differing from the original only slightly in quality but significantly in price. It is not known to what degree, if at all, the mentioned calculation of savings is reflected in the official index of the cost of living.
44. B. Csikós-Nagy, *Valóság*, no. VI, 1974, and P. Vlach, *Hospodářské noviny*, no. 36, 1968.
45. See also A. I. Katsenelinboigen, 1975.
46. See also P. J. D. Wiles, 1977, p. 375.
47. In the second half of the 1950s Czechoslovakia tried to use price reductions to improve the standard of living of families with children. It took some time to realise that this is not an effective way to do it.
48. G. Grossman (1977, pp. 136–7) mentions that the goal of Soviet price policy is characterised by their authoritative sources as a combination of stability and flexibility (in contrast to frozen prices). For our purpose the terms 'policy of rigid price stability' and 'flexible price stability' (the first referring to the present Soviet, GDR and Czechoslovak system and the second to the Hungarian, Polish and the Czechoslovak of the 1960s) better reflect realities in the examined countries. In contrast to the Soviets who hold down prices even at the cost of market disequilibria, the Hun-

garians, though striving for price stability, nevertheless use prices for avoiding disequilibria.
49. See *Neues Deutschland*, 2 June 1962.
50. See *Monitornii obzor SSSR*, no. 801, 4 January 1977; *Neue Zuercher Zeitung*, 12 January 1977.
51. M. Bornstein, 1976, p. 52.
52. For more. see G. Garvy, 1974, p. 88; A. Marton vol. 14(4), 1975, and J. Mujżel, *Ekonomista*, no. 4, 1974.
53. For the way in which enterprises take advantage of the introduction of new products for price increases, see G. Grossman, 1977, pp. 139–42 and F. Haffner, *Zeitschrift für Wirtschafts und Sozialwissenschaften*, no. 2, 1977.
54. According to *Neue Zuercher Zeitung* (12 January 1977) Western experts estimate that price increases in the USSR range from 2–5% annually. G. Schroeder, 1975, estimated repressed inflation at several percentage points annually.
55. See also R. Portes, May 1977.
56. According to G. Schroeder and B. Severin (1976, p. 631), the real increase in the Soviet price index was 29% during 1955–75, of which 10.8% occurred in 1970–5.
57. D. Howard, no. 4, 1976, estimates the increase in prices at 15.1% during 1955–72.
58. *Monitornii Obzor SSSR*, 4 January 1977. The government is, e.g., paying a subsidy of 3.21 roubles per kg beef (its retail price is 1.65 rouble) and 1.34 roubles for each kg butter. According to Constance Krueger (no. 2, 1974) the subsidies in 1971–5 were to have been twice as high as they had been in 1966–70.
59. M. Bornstein, 1976.
60. Interval materials of Radio Liberty, 1977.
61. M. Melzer, *Deutsches Institut für Wirtschaftsforschung*, 1969, and W. Sztyber, *Ekonomista*, no. 6, 1969.
62. In connection with the third stage of the wholesale price reform, prices of some industrial goods were increased. See *Neues Deutschland*, 15 November 1966.
63. *Neues Deutschland*, 19 November 1971.
64. See *Die Zeit*, 12 March 1976.
65. In view of the unfavourable reaction to the 1957 price increases, the Hungarian government promised to refrain from further increases. (See *Gazdaságpolitikánk tapasztalatai és tanulságai 1957–1960*, 1976, p. 346.)
66. Ibid. p. 347.
67. Readers who are eager to know more about this subject are advised to read L. Beskid, *Gospodarka planowa*, no. 11, 1957, and R. Glowacki and Cz. Kos, *Zycie gospodarcze*, 3 April 1960. The former gives a comparison between Polish relative prices and foreign ones, whereas the latter makes a comparison of postwar prices with the prewar ones.
68. There are, of course, exceptions, e.g., sugar.
69. Some Hungarian computations show that the Hungarian price system acts only moderately in the way mentioned. See J. Ladányi, *Valóság*, no. 12, 1975; T. Érsek, *Pénzügyi Szemle*, no. 10, 1972.

70. In Hungary they were fixed in 1951, in Poland and Czechoslovakia in 1953.
71. See B. Csikós-Nagy, *Finance a úvěr*, no. 3, 1968.
72. B. Csikós-Nagy, in *Reform of the Economic Mechanism in Hungary*, 1969, p. 140.
73. The first price increase action designed in such a way was in 1963 (see *Trybuna Ludu*, 15 September 1963). See also J. Struminski, *Gospodarska planowa*, no. 1, 1966. This policy only lasted a short time, however.
74. In Hungary this policy was applied for the first time in 1966.
75. See A. J. Smith, in *East European Economies Post-Helsinki*, 1977, pp. 162–3.
76. See B. Csikós-Nagy, 1971, p. 418. In 1968, 23% of products for private consumption were sold at free market prices. It was expected that in 1975 market prices would reach a 33% share. (See L. Köszegi, no. 2, 1975).
77. B. Csikós-Nagy, *Finance a úvěr*, no. 3, 1968, and also from the same author, in *Reform of the Economic Mechanism in Hungary*, 1969.
78. *Radio Free Europe Research*, 28 December 1965, and 22 January 1966; *Neue Zuercher Zeitung*, 16 January 1966.
79. See A. Marton, *Kereskedelmi Szemle*, no. 12, 1975.
80. B. Csikós-Nagy, *Marketing in Ungarn*, no. 3, 1976. The resolution of the Central Committee of the Hungarian Communist Party of April 1968 (*Társadalmi Szemle*, no. 5, 1978) mentions that, due to subsidies, the level of wholesale prices is at present higher than the level of retail prices.
81. See Z. Fallenbuchl, no. 1–2 1973(b).
82. According to *Trybuna Ludu*, 3 December 1970, the scheduled price increases totalled Zł15.7 milliard, and decreases Zł10.9 milliard.
83. *Radio Free Europe Research*, Poland, 15 January 1971.
84. It would be interesting to analyse the reasons for the different reactions to price adjustments in Poland and in Hungary. As already mentioned, the Hungarians carried out a huge price adjustment in 1966 which was generally accepted with concern, in some circles even with resentment. These feelings were not reflected in outward manifestations as they were in Poland. One of the reasons for the different reaction was the fact that the Hungarians took time to prepare the population for the need of price adjustments and also gave more adequate compensation. What is perhaps no less important is that the Hungarian people have more confidence in their government than the Poles in theirs. (See also V. Zorza, *The Guardian*, 17 December 1970.)
85. *Trybuna Ludu*, 28 February 1971 and *Süddeutsche Zeitung*, 17 February 1971.
86. *Trybuna Ludu*, 9 January 1971.
87. The freeze affected 65% of all the foodstuffs consumed by an average family. (See W. Pruss, 1975, p. 70.)
88. Ibid. p. 71.
89. The scheduled price increases for food, including a 40% increase in meat prices, would have meant a 16% increase in the cost of living. See R. N. Gorski, *Berichte des Bundesinstituts für ostwissenschaftliche und internationale Studien*, no. 36, 1976.

90. *The Financial Times*, 25 June 1976 and *Radio Free Europe Research*, Background report 176, 16 August 1976.
91. See *Stern*, Hamburg, July 1976.
92. P. Vlach, *Hospodářské noviny*, no. 36, 1968.
93. The Economic Guidelines for 1969 and the collective agreement concluded between the government and the Trade Unions were based on this concept. See *Práce a mzda*, no. 11, 1968.
94. Even if average price increases (ratio of receipts from sale of goods and their physical volume) which reflect changes in assortment are considered, the results are not so dramatic as the anti-reformers try to make out. In 1968 average price increases amounted to 3% and in 1969 to 4.7%. (See S. Hejduk, *Finance a úvěr*, no. 5, 1975.)
95. Ibid.
96. The period 1954–7 in Poland is a period of fast growth in consumption and a slowdown in investment growth.
97. All the Polish figures, if not otherwise indicated, are from *Rocznik Statystyczny 1976*, pp. XXXIV–XXXVII and 85.
98. As is known, Poland has had a much higher increment in the population of working age than Hungary and Czechoslovakia. See G. Baldwin, in *East European Economies Post-Helsinki*, 1977, p. 421.
99. W. Krencik, 1977, p. 151.
100. The personal wage fund in the socialist sector is smaller than the non comprehensive wage fund in the national economy due to several items. The difference is approximately 5%. See J. Meller, 1977, p. 120.
101. J. Meller, 1977, pp. 102, 110, 120, 122 and 155.
102. The inquisitive reader who would like to see a more comprehensive analysis is advised to read M. Kucharski, ch. 6, 1972.
103. See also B. Schwarz, in *Struktur- und Stabilitätspolitische Probleme alternativen Wirtschaftssystemen*, 1974, p. 157.
104. See W. Krencik, 1977, p. 232.
105. See J. Meller, 1977, pp. 130 and 150.
106. In the period 1970–5 prices of pork increased by 33% on the free market. (See *Rocznik Statystyczny*, 1976, p. 400.)

NOTES TO CHAPTER 2

1. See, e.g., K. Laski, 1967, p. 194.
2. By 'average incomes' is meant all incomes regardless of source. In this chapter average wages refer to the whole economy.
3. Having in mind such a situation some economists suggest linking average wage growth with production growth in consumer goods industries. See M. Toms and J. Fogl, *Hospodářské noviny*, no. 36, 1968.
4. We have mentioned several factors which cause inflation even if wages grow at the same rate as productivity. M. Kucharski (ch. 6, 1972 and also *Finance a úvěr*, no. 5, 1968) discusses this problem indirectly when he examines in a systematic way the limits for real wage increases per employee.

5. The application of this relationship is due to the Marxist concept of national income accounting.
6. B. Levčik, 1969, p. 32.
7. See, e.g., A. Zverev, *Voprosy ekonomiky*, no. 11, 1962; *Ökonomik der Arbeit in der DDR*, 1963; M. Derco and M. Kotlaba, *Plánované hospodářství*, no. 12, 1976 and W. Krencik, *Gospodarka planowa*, no. 6, 1976.
8. The relationship that existed between growth of wages and productivity and which was conditioned by certain circumstances, especially for some time by the growing investment ratio, which was a consequence of rapid industrialisation and low investment returns, and also by shortages of consumer goods, was raised to the status of a general law. It has become a dogma (which for a long time it was not advisable to question) like another related dogma—that a faster growth of production of producer goods than consumer goods is a necessary condition for economic growth.
9. In comparing wages to productivity, caution is warranted when choosing the method of calculating productivity. Naturally an effort must be made to choose the method which best reflects real changes in productivity. It seems to us that for our purposes national income per worker employed, which is also a productivity indicator, meets the criterion relatively well. This could be challenged by a reference to the shortcomings of the Marxist concept of national income accounting, mainly due to the fact that it does not include services. However, if there is no difference between growth rates in the material and non-material spheres, Marxist and Western concepts of national income accounting should arrive at approximately the same estimates of growth. (See S. Cohn, pp. 126–7 and 136.) Since our interest is really concentrated on growth rates in national income, we have refrained from the time-consuming recalculation of the Soviet bloc countries' figures, all the more because it is a justified assumption that not all the figures needed for such a recalculation are available for all the countries under review. The method of calculation of productivity used, that is, per worker, is not an ideal solution. However, the calculation of productivity in terms of man-hours is impossible for lack of adequate figures.
10. As mentioned, employees of the army, security branches, the party apparatus, and some other institutions are not usually included in these figures.
11. In both productivity indicators collective and private farmers are not included in the denominator. Disregarding the fact that their output is included in national income, such an approach could be justified by arguing that collective farmers are not employees and thus not wage recipients. Yet not including collective and private farmers in the denominator in a sense distorts the productivity indicator. In all the countries in the first half of the 1950s, and in some later too, employment in the non-agricultural sectors grew fast, considerably reducing the rate of growth of national income per employee compared to aggregate national income. To a great degree the fast increase in employment was at the expense of agriculture. This means that for the economy as a whole (including collective and private farmers) the growth rates in productivity (national income per earner) are higher than SLP and NIE. Yet if produc-

tivity is computed in this way, it is only logical that it must be compared with its counterpart—average earnings (average wages + average incomes of collective farmers) instead of average wages. And, as already mentioned, the focus of the study is on wages and not on average earnings. In addition figures on average earnings are available only for the period starting with 1960 (Czechoslovakian figures go back to 1955) and not for all countries. A glance at statistical figures shows, as could be expected, that average earnings from the 1960s on grew faster than average wages, due to the fact that incomes of collective farmers grew faster than wages in the state sector. In the USSR incomes of collective farmers derived from work in collective farms including the equivalent for rewards in kind increased by 222.5% in the period 1960–75, whereas average wages increased by only 80.8%. (See *Narodnoe khoziaistvo 1922–72*, p. 243, and *1975*, pp. 414 and 531.) In Czechoslovakia, the differences are not so dramatic but they are nevertheless substantial. The increase in incomes of collective farmers is so great that it more than offsets the impact of the declining share of collective farmers in the total number of earners and the fact that incomes of collective farmers are still smaller than average wages. (In our computation, earners consist of workers and employees and collective farmers. We have left out the private sector since figures for it are not available.) If, however, the fact is taken into consideration that national income per earner grows faster than NIE and SLP, then there is no substantial difference in the long-run between the comparison of average wages with national income per employee and average earnings with national income per earner. If the 1950s could be included, the picture would be somewhat different. Since at that time average wages grew faster than incomes of collective farmers, it can be assumed that average earnings developed less favourably than average wages.

12. For more, see *Economic Survey of Europe 1969*, part I, pp. 5–6.
13. See pp. 10–11. This means that if these one-time wages increases are deducted, the average wage increase will drop to 7.1% in Hungary and to 6.8% in Poland.
14. See pp. 10–11.
15. 'While the national income in constant prices increased in 1948–50 approximately by 20%, the share of population consumption in national income increased by 6%, accumulation by more than 54%. In 1951, funds earmarked for accumulation increased again by 70%.' (See T. I. Berend, 1974, p. 78.) There was a similar situation in Poland. See M. Kucharski, 1972, p. 315.
16. E.g., in Czechoslovakia within the framework of the triple increase of production in the engineering industry, production for military purposes increased the most rapidly (it rose sevenfold). The production of machine-tools and steam and water turbines also increased considerably. Conversely, output of the engineering industry, earmarked for private consumption, decreased; (the production of passenger cars dropped by 70%, that of tractors by 38%, the production of shoemaking machines and machines for the leather industry fell by 72%.) (See V. Lukeš, K. Rybníkář, *Plánované hospodářství*, no. 12, 1968).

17. E.g., in Poland the six-year plan (1949–55) in agriculture was fulfilled only to 75%. See J. Meller, 1977, p. 64. According to M. Kucharski (1972, p. 314) agricultural output increased by 13% in the period 1949–55, whereas final industrial output increased by 110%.
18. 'In examining the efficiency of investment the Party leadership in 1951 concluded unanimously according to confidential minutes: "Investments are always determined by political aspects and economic indicators should play only a secondary role".' (See T. I. Berend, 1974, p. 80).
19. According to J. Meller (1977, p. 6) employment increased in the period 1949–55 according to the plan targets. However, 20% of the newly employed (500,000) were redundant from an economic viewpoint.
20. We disregard black market prices which are not even included in the figures indicated on price increases.
21. This is a payment form according to which over-fulfilment of work norms is rewarded progressively.
22. Real wages increased in this period by 16.1% in Czechoslovakia, 28% in the GDR, 22% in Hungary, 19% in Poland and 6% in the USSR.
23. M. Kucharski (1972, p. 342) maintains that the wage increases were also due to the desire to make changes in wage differentials.
24. J. Montias, 1964, pp. 240–1.
25. See also W. Krencik, 1977, pp. 142–6.
26. According to V. F. Maier (1963, pp. 166–7) wages in the coal industry where wage rates were raised on the average by 56%, increased by 26%. In other branches increases were modest.
27. In Czechoslovakia, which witnessed an absolute decline in economic growth, the slowdown was largely caused by a drop in agricultural output. (See *Economic Survey of Europe in 1969*, p. 50).
28. See B. Bukhanevich, *Voprosy ekonomiki*, no. 8, 1972.
29. W. Krencik (1972, pp. 168–9) maintains that in the years 1958–68 average wages might have increased, due only to changes in qualification mix by 1% annually.
30. The mentioned increase in average wages was largely a result of increases in the minimum wage and in wage rates in some branches of industry. According to R. Batkaev (*Sotsialisticheskii trud*, no. 3, 1971) increases in wage rates accounted for one-third of the average wage increases in industry.
31. B. M. Sucharevskii, in *Trud i zarabotnaia plata v SSSR*, 1975, p. 229.
32. In that period the SLP grew more slowly than NIE in Poland and Czechoslovakia. As for Poland, there is a good explanation; the SLP is computed from national income produced which grew by 9.8% annually on the average; whereas the NIE is computed from national income distributed which, due to the great deficit in foreign trade, grew much faster (12%).
33. See also J. Meller, 1977, p. 125.
34. Ibid., p. 150, and also *Economic Survey of Europe in 1971*, part II, p. 115.
35. See W. Krencik, 1977, p. 164.
36. See *Economic Survey of Europe in 1975*, p. 127.
37. One could argue that even NIE is biased on the high side. If services were

included in national income, the new concept, as computations show, would lower the growth index, though not much. See *Economic Survey of Europe in 1969*, pp. 7–8.
38. See J. Meller, 1977, p. 150.
39. The planners envisaged an increase in national income for 1966–70 of 6.6–7.1% annually, but the actual increase was 7.1%. (See *Economic Survey of Europe in 1970*, part II, pp. 67–8.)
40. National income was supposed to grow by 7.3% annually according to the seven-year plan 1959–65, and it grew by 6.5%. For 1971–5, the figures were 6.8% and 5.7% respectively. See *Economic Survey in Europe in 1966*, p. 51; *1971*, part II, pp. 67–8; *1974*, part I, pp. 68–9, and *Narodnoe khoziaistvo SSSR za 60 let*, p. 485.
41. According to M. Kabaj (*Gospodarka planowa*, no. 2, 1978), in the period 1961–76 the growth of real wages in Polish industry amounted to 56% of productivity growth; (if the 1971–6 period of rapid wage increases is disregarded, then the coefficient was only 33%). On the other hand, in 1960–72 in twelve advanced capitalist countries, the growth of real wages was 91% of productivity growth.
42. The Czechoslovak reformers calculated that in order to make wages an incentive they must grow by at least 4% annually. (Working paper of the Committee for the preparation of the reform).
43. See p. 154.
44. See B. Fick, 1965, p. 312.
45. For more about the USSR see J. S. Berliner, 1976, p. 165.
46. See J. Zielinski, 1973, p. 260.
47. See also P. J. D. Wiles, 1977, p. 372.
48. See O. Kiss and V. Havránek, *Finance a úvěr*, no. 11, 1975.

NOTES TO CHAPTER 3

1. Planned economies of the Soviet type are also interested in regulating enterprises' expenditures on social and cultural amenities and on housing for their employees. This is, however, entirely beyond the scope of our study.
2. S. Shkurko, *Sotsialisticheskii trud*, no. 1, 1975.
3. Recently the USSR reduced the number of grades to six and Czechoslovakia introduced a nine grade system.
4. In the USSR, for example, in 1972, 56.8% of manual workers were pieceworkers, but in 1956 there were even more, 77.5%. (See S. I. Shkurko, 1975, pp. 100, 102.)
5. In the USSR the last adjustment of wage rates was in 1969–75 and before that in 1956–65 (1956–60 in productive and 1964–5 in non-productive industries). See B. M. Sucharevskii, in *Trud i zarabotnaia plata v SSSR*, 1975, p. 249. In Czechoslovakia the last adjustment was in 1973, and before that in 1958–60; in Poland in 1972–6, before that in 1960; in Hungary in 1976, and before that in 1971; the GDR is now engaged in adjustment and was before in 1958–60. See B. Sucharevskii, *Sotsialisticheskii trud*, no. 4, 1974.

6. S. Shkurko, 1975, pp. 100-2. See also V. P. Gruzinov, 1968, p. 88, for figures on other countries.
7. S. Shkurko, 1975, p. 102.
8. E. I. Kapustin, in *Trud i zarabotnaia plata v SSSR*, 1975, pp. 280-1; Supplement to *Hospodářské noviny*, no. 4, 1974; and B. Sucharevskii, *Sotsialisticheskii trud*, no. 4, 1974.
9. E. I. Kapustin, ibid., pp. 281-2 and V. Maier, 1977, pp. 183-5.
10. Supplement to *Hospodářské noviny*, no. 4, 1974.
11. J. Berényi, 1974, p. 26.
12. W. Krencik, 1972, pp. 78-81.
13. J. Berényi, 1974, pp. 29 and 107.
14. The great difference between the lower and upper limits of wage rates in individual sub-groups together with the bonus system contributed to the widening of wage differentials and to the dissatisfaction of many workers. Therefore, in 1974, the government supplemented the existing system by introducing wage rates for the 110 most important trades in industry and construction. The new rates range only within limits of 30%. (See *Magyar Közlöny*, 22 March 1974, and *Népszava*, 27 September 1974.)
15. The increase should be differentiated; wage rates of sub-groups marked with higher skills and more difficult working conditions should increase by a greater rate, whereas the range within the sub-group should be narrowed. See A. Szávai, *Munkaügyi Szemle*, no. 1, 1977, and *Munka*, no. 2, 1977, and no. 1, 1978.
16. J. Bury et al., 1976, p. 71.
17. Naturally, in assigning the wage-bill the planners must take into consideration the planned growth of average wages. On the other hand, the regulation of the wage-bill allows central authorities indirect control of average wages.
18. What happened in Poland in the 1960s is a good example. See B. Fick, *Zycie gospodarcze*, no. 1, 1964. T. Kierczyński and U. Wojciechowska, (1972, pp. 126-7) drew the following conclusion from the Polish example. 'The direct limitation of the wage-bill (from the centre) led to a concealment of reserves in labour productivity in the economy and also to lack of enterprises' interest in a rational economising on labour force.'
19. This is also one of the reasons why the Shchekino experiment in the USSR (the essence of which is that enterprises receive a portion of the money savings which result from labour saving) has remained limited to a relatively small number of enterprises.
20. For example, the USSR (see *Khoziaistvennaia reforma v USSR*, 1969) and Poland before 1973 (see *Zycie gospodarcze*, no. 11, 1972).
21. A case in point was the Hungarian experience of 1968-70. One of the reasons for resorting to average wage regulation in 1968 was the fear that wage-bill regulation would lead to unemployment. (See J. Lökkös, *Közgazdasági Szemle*, no. 2, 1976.) In fact, just the opposite happened. For more, see Chapter 9.
22. For example, the Czechoslovak reform, as will be shown later, tried to regulate employment by imposing a tax on enterprises which expanded employment.
23. J. Wilczek, *Közgazdasági Szemle*, no. 1, 1972.

24. 'The much-debated average wage regulation has the undeniable advantage'—concludes the Hungarian *Figyelö* (no. 18, 1978) from a panel discussion on wage regulation—'that it is suitable for the regulation of the purchasing power'.
25. For more, see Chapter 9.
26. Fore more, see Chapter 10.
27. J. Lökkös, *Közgazdasági Szemle*, no. 2, 1978.
28. References to statements in this subchapter which will be discussed in greater detail in Part Three are given there.
29. Gross value of output is computed in constant wholesale prices and consists primarily of the value of final products, semi-finished products and subcontracting services performed for other enterprises. It also includes changes in the value of inventories of semi-finished products produced within the enterprise for their own purpose.
30. The terms success and evaluation indicators are used as synonyms.
31. In the recently reformed Polish system, though planning has become more flexible and less directive, binding targets from the centre are retained. Apart from the mandatory normatives connected with wage regulations and incentives, the appropriate minister is authorized to assign enterprises the following targets *inter alia*: value of the shipments of goods for the market, introduction of new products, value of exports and imports broken down according to geographical areas and size of investment credits. See J. Marzec in *Nowy system ekonomiczno finansowy w organizacjach przemysłowych*, 1974, p. 79; and J. Olszewski, *Nowe drogi*, no. 4, 1974.
32. For more, see Chapter 7.
33. For more about the role of the monobank, see G. Garvy, 1966, pp. 28-42.
34. See also J. S. Berliner, 1957, pp. 282-3.
35. See *Gesetzblatt der Deutschen Demokratischen Republik*, no. 5, 1972.
36. *Biulletin'*, no. 4, 1977.
37. A. Topiński, in *Zarys systemu funkcjowania przemysłowych jednostek inicjugcych*, 1975, p. 149.
38. See *Sbírka zákonů*, announcement no. 15, 1975.
39. In Poland, collective agreements are concluded between the directors of central boards (ministers) and the chief board of the TU. See *Diennik ustav*, no. 47, 1974.
40. See, e.g., P. Wiles, 1974, pp. 131-2.
41. R. Portes, May 1977.
42. See *Népszava*, 25 February 1976.
43. *Sbírka zákonů ČSSR*, no. 157, 1975.
44. See also M. McAuley, 1969, p. 83.
45. *Sbírka zákonů ČSSR*, no. 63, 1966.
46. Attempts to confine collective agreements to the material interests of the workers are denounced by all the Communist parties. See, e.g., *Život strany*, no. 3, 1972.
47. L. Zsóka, *Munkaügyi Szemle*, no. 11, 1976.
48. See *Népszava*, 25 February 1976, and *Odborář*, no. 5, 1976.
49. See *Odborář*, no. 5, 1976.

NOTES TO CHAPTER 4

1. In all the countries, bonuses are also granted from the wage-bill. In some countries even engineering-technical personnel receive bonuses from the wage-bill.
2. The term bonus will be used generally in lieu of all other rewards above basic wages. The terms 'bonus fund' and 'incentive fund' are used as synonyms.
3. Of course, the share of bonuses in employment incomes of individual groups of employees differ. In top managers' incomes bonuses have the biggest weight.
4. The Polish reform of 1973 has attempted to make the regulation of basic wages as the main incentive. (See A. Topinski, in *Zarys systemu funkcjonowania* . . . 1975, p. 142.
5. See P. Bánki, *Közgazdasági Szemle*, no. 3, 1965.
6. Under the term 'top managers' we have in mind directors of enterprises and their deputies. The term is not used in the same way in all the countries. The term 'managerial staff', which is used in connection with the Polish reform, is much broader. It includes all managerial staff down to supervisors and experts.
7. Poland did not take part in the wave of reforms of the second half of the 1960s. It carried out a major reform in 1973 which, unlike the reforms in other countries, did not include the creation of a separate bonus fund for all employees.
8. Nowadays in all the countries enterprises have, apart from the bonus fund, a fund for social and cultural needs and housing (which is termed differently in different countries). This fund is beyond the scope of our study.
9. One qualification must be made. The statement is true insofar as it refers to the period of reforms in the 1960s when all the countries tried to make the system of management more conducive to incentives. In making the decision whether the bonus fund should be separate or not, the planners surely must have considered this question in the context of the prevailing system of management.
10. E.g., J. Krizsanits, *Munka*, no. 2, 1972.
11. For more, see Chapter 5.
12. Commodity production differs from gross value of output in that it is confined to goods earmarked for selling. It should not be mixed up with sales which are an indicator of the goods sold. For more see J. Jilek et al., 1976, p. 88.
13. *Ekonomicheskaia gazeta*, no. 22, 1971 and no. 23, 1972. See also V. Ignatushkin, *Finansy SSSR*, no. 3, 1973; G. Schroeder, 1973, pp. 31–5, and V. A. Rzheshevskùù, *Dengi i kredit*, no. 10, 1975.
14. A. J. Miliukov, 1977, p. 112. In the official document on the formation of the incentive fund, there is no mention of Miliukov's assertion. See *Biulletin*, no. 4, 1977.
15. *Ekonomicheskaia gazeta*, no. 23, 1972, and P. Bunich, *Ekonomika i organizaciia promyshlennogo proizvodstva*, no. 4, 1975.
16. *Biulletin'*, no. 4, 1977.

17. V. A. Rzheshevskii, *Planovoe khoziaistvo*, no. 3, 1973 and *Biulletin*, no. 4, 1977.
18. *Gesetzblatt der DDR*, part II, no. 16, 1971.
19. In 1972 for an increase of one per cent in commodity production, the bonus fund could increase by $1\frac{1}{2}\%$, whereas for profit the normative was $\frac{1}{4}\%$. In 1974 the normative was increased.
20. See *Gesetzblatt der DDR*, part II, no. 5, 1972.
21. *System und Entwicklung der DDR-Wirtschaft*, 1974, pp. 78–9.
22. A. Suchá and V. Wosková, *Práce a mzda*, no. 10, 1978.
23. Authorities may assign another indicator if profit is not suitable for the enterprise in question.
24. This is mainly in case enterprises are planning losses or expect fluctuations in profit during the year, and, finally, if the bonus fund calculated by the first method would amount to more than 30% of the profit. See *Sbírka zákonů ČSSR*, no. 158, 1970; no. 165, 1971 and no. 157, 1975.
25. See M. Majtan, *Finance a úvěr*, no. 11, 1973 and E. Moravec, *Práce a mzda*, no. 7, 1973.
26. E. Moravec, *Práce a mzda*, no. 3, 1974.
27. *Sbírka zákonů ČSSR*, no. 157, 1975.
28. Figures on the bonuses as a percentage share of profit in Soviet industry show that there is a small correlation between both. See Iu. Artemov, *Voprosy ekonomiki*, no. 5, 1975.
29. For more, see Chapter 7.
30. *Gesetzblatt der DDR*, part II, no. 67, 1968, and K. P. Hensel, in *Wirtschafts-systeme des Sozialismus im Experiment—Plan oder Markt*, 1972, pp. 155–60.
31. For more, see Chapter 9.
32. *Khoziaistvennaia reforma v SSSR*, 1969, pp. 227–30; *Ekonomicheskaia gazeta*, no. 23, 1972, and *Biulletin'*, no. 12, 1972.
33. *Sotsialisticheskii trud*, no. 9, 1977.
34. See *Práce a mzda*, no. 3, 1971.
35. See *Hospodářské noviny*, no. 18, 1976.
36. See *Gesetzblatt der DDR*, part II, no. 52, 1971 and part II, no. 34, 1972.
37. It is worthwhile mentioning that the instructions for awarding directors of associations also list as criteria creation of scientific-technical preconditions for the development of the associations and their involvement in the improvement of working and living conditions of the workers. See *Gesetzblatt der DDR*, part II, no. 52, 1971.
38. D. Granick, 1974, p. 234.
39. *Monitor Polski*, no. 56, 1972.
40. *A gazdasági szabályozó* . . . , 1975, pp. 74–82; A. Szávai, *Munkaügyi Szemle*, no. 12, 1975.
41. For more about the incentive system see J. Adam, *Revue d'études comparative est-ouest*, June 1977.
42. According to ministerial regulations from 1971, which seem to be in force even now, top managers can receive only the annual reward, which is from the bonus fund. The only exceptions are rewards for inventions, patents, and rewards at personal jubilees. See *Práce a mzda*, no. 3, 1971.
43. See Iu. Margulis, *Finansy*, no. 7, 1973.

44. A. Topiński, in *Zarys systemu* . . . , 1975, p. 143.
45. D. Granick, 1974, p. 234.
46. In Czechoslovakia it seems to be included in the annual reward. Top managers are not allowed to receive a year-end reward as a separate bonus. See *Práce a mzda*, no. 3, 1971.
47. Iu. Artemov, *Voprosy ekonomiki*, no. 5, 1975.
48. For more about the evolution of bonuses see J. S. Berliner, 1976, pp. 478-9.
49. A. Miliukov, 1977, p. 107.
50. Iu. Artemov, *Voprosy ekonomiki*, no. 5, 1975.
51. *Sotsialisticheskii trud*, no. 9, 1977.
52. In the past there was a ceiling of a four month's salary; an additional two months' basic salary was allowed for the introduction of new technology. (See G. Schroeder, 1973, p. 35; Iu. Margulis, *Finansy*, no. 7, 1973).
53. E. Moravec, *Práce a mzda*, no. 12, 1974.
54. The Czechoslovak government plans for a much higher bonus fund for the period 1976-80. (See Y. Cima, *Plánované hospodářství*, no. 5, 1976.)
55. *Statistická ročenka ČSSR*, 1975, p. 264.
56. A. Baloušek, *Práce a mzda*, no. 4-5, 1976.
57. *Rocznik statystyczny 1976*, p. 166.
58. Z. Jacukowicz, 1974, pp. 180-9.
59. *Statisztikai Évkönyv 1970*, p. 166.
60. *Statisztikai Évkönyv 1975*, p. 156.
61. *Bér- és jövedelemarányok az iparban, 1967-71*, 1973, pp. 37, 38, 54; *Foglalkoztatottság és kereseti arányok, 1973*, pp. 161-2.
62. *A gazdasági*, see pp. 78-81 and 230.
63. See *Foglalkoztatottság és keresti arányok*, 1973, pp. 161-2, 173, 191, and J. Adam, *Revue d'études comparatives est-ouest*, June 1977.
64. *Gesetzblatt der DDR*, part II, no. 5, 1972.
65. A. Suchá and V. Wosková maintain (*Práce a mzda*, no. 10, 1977) that DM900 was also fixed for the period 1976-80.
66. According to D. Granick (1975, p. 197), who studied the East German system on the spot, year-end rewards made up 66-85% of the bonus fund in the enterprises which he interviewed.
67. *Presse-Informationen*, 18 April 1974.
68. D. Granick indicates (1975, p. 200) figures for three enterprises in 1969. According to them the differences in bonuses were large; in one they were 91% of the month's salary, in the second 236%, in the third 550%.

NOTES TO CHAPTER 5

1. To him an enterprise is—to put it generally—the more efficient the less difficulties it has in paying taxes.
2. M. Sokol, *Plánované hospodářství*, no. 2, 1968.
3. In his paper, *Plánované hospodářství*, no. 3, 1965, Sokol analysed the pros and cons of the principle of a uniform tax on gross income.
4. M. Sokol, *Plánované hospodářství*, no. 2, 1968.
5. Ibid. See also M. Sokol, *Plánované hospodářství*, no. 8-9, 1966.

6. Z. Kodet, *Politická ekonomie*, no. 4, 1965.
7. Ibid.
8. J. Kosta and B. Levcik, *Österreichische Osthefte*, no. 6, 1967. See also B. Fick, 1965, p. 313.
9. J. Pajestka, *Życie gospodarcze*, no. 27, 1973.
10. See also U. Wojciechowska, *Życie gospodarcze*, no. 27, 1973.
11. Such a situation really developed in Hungary in the period 1968–70. See J. Adam, November 1974(b).
12. This way of arguing was used in a conference on financial measures held in Prague in August 1968 in which the author took part.
13. A. Bakhurin and A. Pervukhin, *Voprosy ekonomiki*, no. 9, 1963.
14. A. Zverev, *Voprosy ekonomiki*, no. 11, 1962.
15. B. Sulyok, 1969, p. 167.
16. Because, in the eyes of many communist leaders, the Yugoslav system, particularly the workers' councils, bore the hallmark of heresy, it was not wise to state these views publicly.
17. B. Csikós-Nagy, Winter 1965.
18. It seems that this argument is taken from an official document. It is given with the same wording by two authors in two different publications. See B. Sulyok, 1969 and also T. Bácskai, 1971.
19. M. Timár, 1973, pp. 56–57.
20. See Ibid. Similar views surfaced recently. See *Figyelö*, no. 18, 1978.
21. See Roubal and Šourek, in *Řízení národniho hospodářství*, 1967, pp. 163–4.
22. E. Seifert argued (*Wissenschaftliche Zeitschrift der Hochschule für Ökonomie*, no. 2, 1966) in a similar way.
23. H. Nick, *Wirtschaftswissenschaft*, no. 11, 1966.
24. V. Sitnin, 1974, p. 144.
25. See U. Wagner, 1972, p. 66.
26. A. Timár, 1973, p. 50.
27. To mention some non-Soviet economists who argued along these lines: Roubal and Šourek, in *Řízení národniho hospodářství*, 1967, pp. 163–4; H. Fiszel, *Życie gospodarcze*, no. 17, 1973; H. Nick, *Wirtschaftswissenschaft*, no. 11, 1966 and T. Kierczynski and U. Wojciechowska, 1972, p. 43.
28. E. G. Liberman, *Pravda* 9 September, 1962; also *Soviet Life*, July 1965.
29. V. S. Nemchinov, *Pravda* 21 September 1962.
30. B. Ward, 1958.
31. B. Csikós-Nagy, 1972.
32. B. Sulyok and also T. Bácskai, see Note 18. Again, the wording is the same in both publications.
33. See also P. Ernst, *Plánované hospodářství*, no. 8, 1969.
34. See also U. Wojciechowska, *Życie gospodarcze*, no. 27, 1973.
35. For more details see T. Kierczyński and U. Wojciechowska, 1972, pp. 43–98.
36. See the Resolution of the Central Committee of the Hungarian Communist Party, *Társadalmi Szemle*, no. 5, 1978, p. 11.
37. See *Finanse*, no. 2, 1970.
38. E. Liberman, *Ekonomicheskaia gazeta*, no. 51, 1965.
39. By net indicator is meant an indicator which includes only the enterprises'

own performance; this means it is net of material costs at least. In this sense gross income is a net indicator.
40. V. Sitnin (1974, p. 145) maintains that the Czechoslovak model of 1966–9 lacked logical coherence by not making gross income a genuine sales indicator.
41. For more, see pp. 138–9.
42. To define performance in such a way in a direct system means that the authorities, on the one hand, assign targets on equitable terms and, on the other hand, purge performance of contributions which are not the work of the enterprise in question.
43. See also B. Fick, 1965, pp. 317–19, and J. Zielinski, 1973, p. 262.
44. Kierczyński and Wojciechowska (1972, pp. 125–31) suggest applying profit as an indicator not only for variable wages (bonuses) but also for the wage-bill (basic wages), arguing that two separate indicators generate conflicts. At a closer look, it turns out that what the authors suggest is linking wages to net output as a substitute for profit and bonuses to profit.
45. V. Sitnin, 1974, p. 147.
46. J. Wilczek, *Közgazdasági Szemle*, no. 1, 1972.
47. See also P. Bunich, *Voprosy ekonomiki*, no. 9, 1975.
48. *Ökonomik der Arbeit*, 1974, pp. 503–5.
49. Czechoslovakia (in 1959–62) was the first country to use a productivity indicator as a sole wage growth regulator. (For more, see Chapter 10).
50. For more, see Part Three.
51. D. Karpukhin and V. Rozhkova, *Planovoe khoziaistvo*, no. 5, 1976.

NOTES TO CHAPTER 6

1. T. Buda and L. Pongrácz, 1968, p. 27.
2. J. Bury *et al.*, 1976, p. 32.
3. Its rates and progressivity were mild, and for this reason it could not be a real impediment for wage increases. See W. Jaworski *et al.*, 1977, p. 207, and E. Moravec, *Práce a mzda*, no. 2, 1978.
4. See, e.g., J. Valach, 1972, p. 143.
5. See *Sbírka zákonů*, Announcement no. 157, 1975, as published in *Práce a mzda*, no. 2, 1976, p. 63.
6. According to V. Sitnin (1974, p. 1972) the contributions of enterprises to social insurance were the following: 17% in Hungary, 10% in the GDR, 15.5% in Poland, 7% in the USSR and 25% in Czechoslovakia; in all cases computed as a percentage of the wage-bill. See also D. Butakov, 1973, p. 172.
7. See Supplement to *Hospodářské noviny*, no. 42, 1975.
8. V. Sitnin, 1974, pp. 166–8 and 189, and A. Nove, 1977, p. 233.
9. In our understanding, 'uniform' means that enterprises pay the same tax rates on taxable wage increases.
10. There are, of course, other possible solutions. See also F. Kölgyesi and A. Timár, *Pénzügyi szemle*, no. 11, 1974.
11. Of course taxation has also to dampen wage differenti

12. At present in Hungary the tax is differentiated only by the forms of wage regulations; within the individual forms, it is uniform. (See Chapter 9).
13. For more, see Chapters 5 and 10.
14. See D. Granick, 1975, p. 268; R. D. Portes, May 1970, pp. 307–13. Y. Kotowitz, R. Portes, no. 3, 1974 and A. Timár, *Munkaügyi Szemle*, no. 1, 1979.
15. For more, see Chapter 10.
16. 'This taxation system,' writes B. Csikós-Nagy (*New Hungarian Quarterly*, Autumn 1974), 'not only proved to be an efficient anti-inflationary method but in certain sections of production it represented such a severe limitation that a wage increase corresponding to the average raising of wages or representing at least a minimum wage improvement had to be made possible through tax preferences.' (Quoted according to Radio Free Europe, 16 September 1974.)
17. It should be made clear that the normative need not always be expressed as a fraction of the fund (wage-bill or bonus fund) of which it is a component. For example up to 1976 in the Soviet incentive system, the normative for the fulfilment of success indicators was expressed in terms of a fraction of the wage-bill.
18. D. Granick (1975, p. 304), who studied the working of the Hungarian system on the spot, found that it does not work entirely as would be expected on the basis of what is known about the principles of the reform.
19. See *Biulletin*, no. 4, 1977, p. 16.
20. K. Golinowski and J. Kalisiak, *Gospodarka planowa*, no. 7–8, 1975.
21. V. A. Rzheshevskii, in *Teoretické základy a praxe hospodářské reformy v SSSR*, 1973, pp. 152–7.
22. See A. Timár, *Munkaügyi Szemle*, no. 12, 1975, and T. Sawczuk, *Gospodarka planowa*, no. 3, 1977.
23. B. M. Sukharevskii, in *Trud i zarabotnaia plata v SSSR*, 1975, p. 252.
24. G. A. Egiazarian, 1976, pp. 158–9.
25. See N. Garetovskii, *Voprosy ekonomiki*, no. 5, 1971.
26. G. A. Egiazarian (1976, p. 158) contends that authorities fix high normatives for poorly performing enterprises and low for those that perform well, an approach which defeats the whole purpose of normatives.

NOTES TO CHAPTER 7

1. F. Holzman mentions that in the period 1933–5 the Soviets had already used this system temporarily (1962, p. 34).
2. V. F. Popov, 1957, p. 167.
3. *Khoziaistvennaia reforma v SSSR*, 1969, p. 23.
4. See *Ekonomicheskaia gazeta*, no. 7, 1972.
5. V. Sitnin, 1974, p. 146, and A. Miliukov, *Sotsialisticheskii trud*, no. 5, 1969.
6. M. Kabaj (*International Labour Review*, no. 2, 1966) contends that in 1945–55 in some branches the adjustment coefficient for over-fulfilment was 2–3.
7. According to Iu. Margulis (*Finansy SSSR*, no. 3, 1976) the adjustment

coefficient ranges from 0.6–0.9, and the different socio-economic groups do not share evenly in the increase in the wage-bill due to an over-fulfilment of output targets. Wages of piece-workers increase, of course, proportionally to increases in output, whereas wage increases of other groups are much smaller. See also V. Maier, 1963, p. 242.
8. P. F. Petrochenko *et al.*, 1965, p. 217.
9. V. Maier, 1963, pp. 242–3.
10. Iu. Margulis, *Finansy SSSR*, no. 3, 1976.
11. True, the central authorities assign output targets, but it is only in some branches of industry that this can be done in great detail for individual products; in most branches the targets have to be aggregated in groups of products and expressed in value units.
12. The fact that in some branches the adjustment coefficient was higher than unity was due mainly to the progressive piece-rates. Before the reform of the wage payment forms (1958–60), 27% of the industrial workers were paid according to progressive piece-rates. See L. J. Kirsch, 1972, p. 25.
13. One important qualification is warranted here. In the 1950s in the USSR, the planners took very much into account in their wage plans the widespread progressive piece-rates and high adjustment coefficients in some branches so that only certain branches witnessed relatively high increases in wages. On the whole wage increases in this period were quite moderate. (See Table 2.2.)
14. See also J. Berényi, 1974, p. 26.
15. See also G. A. Egiazarian, 1976, p. 155.
16. *Khoziaistvennaia reforma v SSSR*, 1969, p. 245, and V. Kletskii and G. Risina, *Planovoe khoziaistvo*, no. 8, 1970.
17. According to E. Liberman, *Ekonomicheskaia gazeta*, no. 5, 1965, sales as an indicator were supposed to be assigned to branches whose products were still in short supply.
18. For more about the reform see R. Campbell, May 1968; M. Ellman, 1969 and J. Adam, October 1973.
19. The plan called for an increase of 33–35% in industrial productivity in 1966–70 (See *Ekonomicheskaia gazeta*, no. 2, 1969) but an increase of only 32% was achieved.
20. According to a study carried out by the Gosplan in cooperation with research institutes, deficiencies in planning and incentives are listed among the reasons for targets not being fulfilled in productivity. (See V. Moskalenko, *Planovoe khoziaistvo*, no. 8, 1971.)
21. See V. Sitnin, *Komunist*, September 1966, and G. Schroeder, *Soviet Studies*, vol. 20, 1968–9.
22. V. Rzheshevskii, *Planovoe khoziaistvo*, no. 9, 1971. The price fixing authorities were hampered in their efforts by a directive that neither consumer prices nor the agricultural sector were to be adversely affected by the revision of wholesale prices. In addition the price authorities had to work with cost calculations of differing degrees of accuracy.
23. V. Rzheshevskii, *Planovoe khoziaistvo*, no. 3, 1973.
24. W. Sielunin, 1971, pp. 109–10.
25. G. A. Egiazarian, 1976, pp. 259–60.
26. For more, see *Biulletin'*, no. 4, 1977.

27. E. Kapustin, *Voprosy ekonomiki*, no. 1, 1976.
28. G. Kiperman, *Ekonomika i organizatsiia promyshlennego proizvodstva*, no. 5, 1976, and J. Dvořák, *Politická ekonomie*, no. 4, 1977.
29. See N. Rogovskii and G. Kiperman, *Voprosy ekonomiki*, no. 2, 1976; J. Dvořák, *Politická ekonomie*, no. 1, 1976 and J. Dvořák, *Politická ekonomie*, no. 4, 1977.
30. E. Kapustin, *Voprosy ekonomiki*, no. 1, 1976; V. Sitnin, 1974, p. 148, and P. Buchin, *Ekonomika i organizatsiia promychlennogo proizvodstva*, no. 5, 1976.
31. N. Rogovskii and G. Kiperman, *Voprosy ekonomiki*, no. 2, 1976, and G. Kiperman, *Ekonomika i organizatsiia promyshlennogo proizvodstva*, no. 5, 1976.
32. P. Khromov, *Voprosy ekonomiki*, no. 5, 1976, and S. Barngolc, *Voprosy ekonomiki*, no. 9, 1976. Barngolc indicates interesting figures on the ratio of profit to the wage-bill. According to him profit was 18 times as high as the wage-bill in the coal industry, 1328 times in oil production, and 127 times in industry as a whole.
33. G. Kiperman, *Ekonomika i organizatsiia promyshlennego proizvodstva*, no. 5, 1976.
34. In conclusion to this survey of the debate, it is worthwhile mentioning that some economists are willing to rehabilitate gross income in the fight for net output. Thus, for example, V. Sitnin (1974, p. 145) argues that it would be wrong to blame gross income for its incorrect application in Czechoslovakia.
35. An example will make clear what we have in mind. Let us suppose that wage costs in one rouble of production are set at 25 copeck, and the value of planned production is 1000 roubles. The wage-bill is in this case $1000 \times 0.25 = 250$.
36. See B. M. Sukharevskii, in *Trud i zarabotnaia plata v SSSR*, 1975, p. 252, and L. Kunelskii, *Sotsialisticheskii trud*, no. 10, 1976.
37. See G. Abramov, *Sotsialisticheskii trud*, no. 6, 1977.
38. The normative coefficient depends on the tautness of the plan, the labour intensity of products, the wage payments forms applied, etc.
39. A. Cernes, *Sotsialisticheskii trud*, no. 12, 1976; G. Abramov, *Sotsialisticheskii trud*, no. 6, 1977.
40. A. Cernes, *Sotsialisticheskii trud*, no. 12, 1976. It seems that in the chemical and machine building industries no difference is made between fulfilment and over-fulfilment.
41. See also Radio Liberty Research, 3 September 1976.
42. V. M. Batyrev, 1947, pp. 165 and 261.
43. Ibid., pp. 262-3. See also F. Holzman, 1962, pp. 37-8.
44. N. D. Barkovskii, *Dengi i kredit*, no. 12, 1976.
45. See V. F. Popov, 1957.
46. *Khoziaistvennaia reforma v SSSR*, 1969, p. 132; see also P. Zadoia, *Dengi i kredit*, no. 1, 1966.
47. See Iu. Margulis, *Finansy SSSR*, no. 3, 1976, and V. S. Zacharov, *Finansy SSSR*, no. 4, 1976.
48. N. D. Barkovskii, *Dengi i kredit*, no. 12, 1976.
49. P. Zadoia, *Dengi i kredit*, no. 1, 1966.

50. *Khoziaistvennaia reforma v SSSR*, 1969, p. 230. This rule was reconfirmed in the recent fundamental principles for awarding bonuses, see *Sotsialisticheskii trud*, no. 9, 1977.
51. See, e.g., V. M. Safirova, in *Finansovo-kreditnii kontrol v narodnom khoziaistve*, 1975, pp. 119–27.
52. P. Zadoia, *Dengi i kredit*, no. 1, 1966.
53. In Hungary, Poland and Czechoslovakia, and to a lesser degree in the USSR, there is a rich literature on this topic. Particularly Hungarian and Polish journals devote continuous attention to wage regulation. By contrast, in the GDR even in journals such as *Arbeit* and *Arbeitsrecht*, or *Sozialistische Finanzwirtschaft* where one would expect that wage regulation would be one of their topics, very little is published on this subject. There is more information in these journals on the USSR and other countries of the Soviet bloc than on East Germany.
54. G. Kohlmey and Ch. Dewey, 1957, p. 554.
55. Ibid., pp. 552–8. *Gesetzblatt der DDR*, 1961, part II, no. 67, p. 450.
56. *Gesetzblatt der DDR*, 1961, part II, no. 67, p. 451.
57. This Party Congress decided to establish an 'economic system of socialism' which was supposed to be a further step in the reform started in 1963. For more see K. P. Hensel, in *Wirtschaftssysteme des Sozialismus im Experiment—Plan oder Markt?*, 1972, pp. 155–69.
58. This is a judgement made on the basis of *Ökonomik der Arbeit* (a quasi-official textbook), 1974, p. 504, and the fact that we have not come across anything in the legislation which would contradict this statement. However, other information available seems to indicate that this was not simply an experiment.
59. See *Lexikon der Wirtschaft*, Arbeit, 1968, p. 421, and T. Krajkovič and V. Sekanina, *Práce a mzda*, no. 10, 1970.
60. See Supplement no. 15 to *Die Wirtschaft*, no. 19, 1970; H. Schönherz and I. Schuster, Supplement 28, *Die Wirtschaft*, no. 32, 1970, and H. Schönherz, *Arbeit und Arbeitsrecht*, no. 20, 1970.
61. *System und Entwicklung der DDR-Wirtschaft*, 1974 and *DDR-Wirtschaft*, 1974, pp. 66–7.
62. K. Hensel, in *Wirtschaftssysteme des Sozialismus im Experiment—Plan oder Markt?*, 1972, pp. 168–9.
63. The legislative provisions which will be noted further on were in force at least up to the end of 1975. The article of A. Suchá and V. Wosková on the German wage regulation system (*Práce a mzda*, no. 10, 1977) allows us to conclude that the present five-year plan has not brought any important changes in the SWR.
64. *Gesetzblatt der DDR*, 1972, part II, no. 10, p. 127.
65. J. Giebner, *Sozialistische Finanzwirtschaft*, no. 1, 1976.
66. *Gesetzblatt der DDR*, 1972, part II, no. 5, p. 57.
67. J. Giebner, *Sozialistische Finanzwirtschaft*, no. 1, 1976.

NOTES TO CHAPTER 8

1. B. Blass, *Gospodarka planowa*, no. 3, 1965.
2. B. Fick, *Życie gospodarcze*, no. 1, 1964.

3. For more, see Chapter 11.
4. B. Blass, *Gospodarka planowa*, no. 2, 1964.
5. For more, see pp. 142–3.
6. See J. P. Farrell, April 1975; B. Fick, *Życie gospodarcze*, no. 1, 1964.
7. The factory fund was small in relation to the planned wage-bill (in 1963 it amounted in industry to 6.4%; see B. Fick, 1967, p. 210), and the changes in it due to a change in financial performance were too small to be an incentive to economise on the wage-bill.
8. To rule on whether or not an over-expenditure was justifiable was a very difficult task for the Bank (see J. P. Farrell, April 1975). It can also be assumed that the Bank was often under pressure to justify the over-expenditure.
9. J. Meller, 1977, p. 85.
10. In the coal, metallurgical and food industries and also in some associations of light industry the automatic system was retained. See B. Fick, *Życie gospodarcze*, no. 1, 1964.
11. The reserve fund was divided into two parts, 'A' and 'B'. 'A' was earmarked exclusively for export purposes and had no limitations.
12. B. Fick, *Życie gospodarcze*, no. 1, 1964; see also *Dengi i kredit*, no. 4, 1964.
13. See B. Fick, *Życie gospodarcze*, no. 2, 1965.
14. J. Meller, 1977, p. 86.
15. See J. Zielinski, 1973, p. 264.
16. See H. Weber, *Życie gospodarcze*, no. 32, 1964.
17. J. P. Farrell, April 1975.
18. B. Krajewski, *Praca i zabezpiezenie spoleczne*, no. 4, 1971.
19. 25% of the bonuses were to be used to reward the fulfilment of the synthetic indicator and 75% for special indicators. See B. Fick, *Finanse*, no. 6, 1970.
20. For more, see ibid., and H. Flakierski, 1973.
21. For 1971 a wage freeze was assumed.
22. B. Fick, *Finanse*, no. 3, 1971.
23. These new associations will gradually replace the old associations. In contrast to the old associations which have predominantly administrative functions, the new associations have economic functions and their performance is based on the principle of '*khozraschet*'. The new associations are directly subordinated to the ministries, and all the assignments resulting from the plan and the principles of the new reform are addressed by the ministries to the associations as a whole. This enables associations to set goals for individual enterprises in a differentiated way. The relation between the new associations and the enterprises integrated in them is not solved uniformly; the relation depends on the type of association. Generally it is possible to state that decisions on development and foreign relations are centralised. The same is true of decisions on drawing investment credit; yet its repayment is left to enterprises. In all types of associations enterprises are autonomous financial units; in a few associations, however, they do not have the position of an independent judiciary unit. For more, see S. Jakubovicz, *Gospodarka planowa*, no. 8, 1973.

24. Z. Fedorowicz, 1977, p. 111; A. Topiński, in *Zarys systemu* ..., 1975, pp. 130–2.
25. See A. Topiński, in *Zarys sytemu* ..., 1975, p. 121.
26. Ibid., pp. 133–5.
27. The initial provisions which applied to associations converting to the system in 1973 were different. For more, see J. Adam, 1975.
28. B. Gliński, *et al.*, 1975, p. 53; J. Bury *et al.*, 1976, p. 54.
29. 'Interim' means that after two years the bonuses may be included in fixed wages, provided enterprises seem to have achieved a permanent improvement in efficiency.
30. For more, see J. Zielinski, 1973, pp. 236–7; B. Fick, 1967, pp. 235–8.
31. This is not to say that every worker has the right to claim an additional monthly wage. The year-end reward of the worker depends on his performance and observation of labour discipline.
32. See Interview with H. Slawecki, *Życie gospodarcze*, no. 50, 1973, and E. Nedzowski, *Życie gospodarcze*, no. 46, 1973.
33. E. Cichowski, *Finanse*, no. 8, 1973.
34. Net profit is the balance of gross profit left after paying off credits and making contributions to research institutes. Gross profit is what remains of financial accumulation (obtained after deducting cost of production from the receipts of output sold and services) after paying the turnover tax, the tax on fixed capital and the tax on the wage-bill (which is 20%).
35. A. Wolowczyk, *Finanse*, no. 5, 1973.
36. To avoid such an expansion of employment, the central authorities also tried to control growth of employment. True, in 1964 the assignment of the number of employed to enterprises was formally abolished. In practice, however, this measure had little effect, partly due to the retention of the directive assignment of wage-bills. See J. Obodowski, *Życie gospodarcze*, no. 11, 1972.
37. Some economists suggested designing the reform so as to create conditions for a new model of high wages combined with an optimal employment. See, e.g., M. Kabaj, *Życie gospodarcze*, no. 19, 1974.
38. A comparison with the Hungarian system has already been given. See Chapter 3.
39. U. Wojciechowska, *Życie gospodarcze*, no. 27, 1973.
40. This also refers to piece-work which many economists would like to see substantially limited or gradually terminated. See, e.g., K. Klocek, *Życie gospodarcze*, no. 33, 1972.
41. Some of the reasons for choosing output added instead of profit were mentioned in Chapter 5. It seems that one of the main reasons was the concern that profit might cause a conflict of interests between management and the other personnel of enterprises. And this is also the reason why a special bonus fund for managerial staff (which is linked to net profit) was introduced. See A. Topiński, in *Zarys systemu* ..., 1975, p. 144.
42. K. Opaliński and I. Zamojska, *Bank i kredyt*, no. 4, 1973.
43. A. Topiński, in *Zarys sytemu* ..., p. 121; K. Golinovski, *Gospodarka planowa*, no. 9, 1977.

44. See, for example, M. Mieszczankowski, *Życie gospodarcze*, no. 17, 1974, and no. 23, 1974.
45. M. Kamiński suggests (*Życie gospodarcze*, no. 25, 1974) checking the present tendency of enterprises to shift to low-taxed products by subjecting the size of the wage-bill to a certain norm of taxation yield. In other words, the wage-bill will be reduced if the yields of the turnover tax drop below the amount corresponding to the norm.
46. See S. Jędrychowski, *Finanse*, no. 4, 1973.
47. See, e.g., A. Zawislak, *Gospodarka planowa*, no. 8, 1973.
48. A. Topiński, in *Zarys systemu* ..., 1975, pp. 129–30, and K. Golinowski, *Gospodarka planowa*, no. 9, 1977.
49. A. Topiński, in *Zarys systemu* ..., 1975, p. 121.
50. B. Holubicki, *Gospodarka planowa*, no. 6, 1977; B. Holubicki, *Gospodarka planowa*, no. 9, 1977; and B. Gliński, 1977, pp. 40–1.
51. B. Gliński *et al.*, 1975, pp. 56–8.
52. See T. Wrzaszczyk, *Gospodarka planowa*, nos. 7–8, 1977.
53. J. Meller, 1977, p. 150.
54. Z. Fedorowicz, 1977, pp. 118–20.
55. It is applied in the following branches: machine, chemical, light and agricultural machinery. See T. Wrzaszczyk, *Gospodarka planowa*, nos. 7–8, 1977.
56. This is the official term for the changes.
57. K. Golinowski, *Gospodarka planowa*, no. 9, 1977.
58. Output added should also be reduced by losses above the norm resulting from poor quality as well as by paid penalties.
59. For more on the modified system see T. Wrzaszczyk, *Gospodarka planowa*, nos. 7–8, 1977, and also *Życie gospodarcze*, no. 21, 1977 and K. Golinowski, *Gospodarka planowa*, no. 9, 1977.
60. B. Csikós-Nagy, 1974, p. 154.
61. *Życie gospodarcze*, no. 21, 1977.
62. The charge for an increase in employment by one person was set at Zł20,000, an amount which corresponded approximately to a six months' average wage in 1974. The charge for wage increase was fixed at 10% of the amount which exceeded the 4% increase in the disposable wage-bill. B. Glinski *et al.*, 1975, p. 59.
63. It seems that no change occurred in the payments for increases in employment.
64. Associations which have not converted to the modified system function according to the pre-1977 provisions.
65. K. Golinowski, *Gospodarka planowa*, no. 9, 1977 and *Życie gospodarcze*, no. 21, 1977.
66. Compared to Hungarian taxes for wage increases above the performance limits, the Polish charges seem to be relatively moderate. For increases of wages above the trigger threshold the usual charges range from 40–96%, probably—the materials available are not specific about this—of the additional wage costs. Individual ministries are allowed to differentiate charge rates by associations. *Życie gospodarcze*, no. 21, 1977.
67. B. Gliński *et al.*, 1975, p. 59.
68. *Życie gospodarcze*, no. 21, 1977.

69. This is a loan extended in cases where enterprises are involved in research or in the introduction of new technology which would be reflected in increased output added in the next year or years. This kind of loan is extended for one year, in exceptional cases for three. (See J. Bury et al., 1976, p. 65.)
70. According to the old provisions, the interest rate was 10% and the loan had to be repaid in the next year. However, if a new overdraft occurred in the year when the repayment was due, the interest rate on the new loan was 12%. (See ibid., p. 63.)
71. Życie gospodarcze, no. 21, 1977.
72. J. Bury et al., 1976, pp. 64–5.
73. K. Golinowski, Gospodarka planowa, no. 9, 1977.

NOTES TO CHAPTER 9

1. I. Buda, Közgazdasági Szemle, no. 4, 1965, and T. Meitner, Közgazdasági Szemle, no. 2, 1958.
2. I. Friss, 1969, p. 37.
3. In 1968, the charge on capital was 5%, the contribution to social insurance 17% and the tax on wages 8%.
4. It should also be mentioned that the reform introduced a production tax to mitigate great differences in profits between enterprises. This tax was levied on profit before its division into two parts. B. Sulyok, 1969, p. 168.
5. I. Buda and L. Pongrácz, 1968, p. 53.
6. Ibid., p. 84. These were averages; within these averages the bonuses of individual managers or wage earners could exceed the fixed limit.
7. Ibid., pp. 94–6.
8. This statement is based on an unpublished study, Budapest, 1972.
9. In order to placate the workers a shift was made to less overt preferential treatment of top managers in the distribution of bonuses. Instead of the single bonus, top managers were given a year-end reward and a profit premium (as a compensation for the smaller year-end reward), both financed from the sharing fund. The former was set in proportion to basic salaries and years of service, and the latter depended on the ratio of the gross sharing fund (before payments for wage increases and taxes were made) to the wage-bill. (See P. Bánki, Közgazdasági Szemle, no. 2, 1973). The gross sharing fund was selected in order not to create conflicting interests between managers and workers. If bonuses of managers had depended on the net sharing fund, managers would have been interested in minimal wage increases. Somewhere along the way the earnings of managers were supplemented by a supplementary profit premium financed from the wage-bill. (The present situation is described in Chapter 4.)
10. In 1968 average wages increased by 2.9%, in 1969 by 5.0% and in 1970 by 6.1%.
11. See F. Flór and P. Horváth, 1972, p. 93.
12. T. Bácskai, CESES, p. 269.
13. Since part-time workers were counted as full-time workers for the pur-

pose of average wage calculation, enterprises showed a special interest in hiring part-time workers. See also D. Granick, 1975, p. 263.
14. A. Kemény, *Práce a mzda*, no. 3, 1971, and L. Pongrácz, *Társadalmi Szemle*, no. 4, 1973.
15. For more on the impact of the system of regulation and incentives see J. Adam, no. 4, 1974.
16. See P. Bánki, *Munkaügyi Szemle*, no. 4, 1971, and B. Balassa, 1973, pp. 352–3.
17. See F. Flór and S. Horváth, 1972, p. 104; L. Kónya, no. 1, 1971.
18. S. Ferge and L. Antal, 1972, p. 66. Statistical figures confirm a substantial increase in the bonus fund. See p. 90.
19. J. Bokor, *Pénzügyi Szemle*, no. 12, 1973.
20. See also P. Bánki, *Munkaügyi Szemle*, no. 4, 1971.
21. F. Flór and S. Horváth, 1972, p. 95.
22. For more, see J. Adam, *Revue d'études comparatives est-ouest*, vol. 8, no. 2, 1977.
23. See the interview with the minister of labour, *Müszaki Elet*, 7 June 1974, and also P. Bánki, *Ipargazdaság*, no. 1, 1974.
24. See p. 63.
25. P. Bánki, *Ipargazdaság*, no. 1, 1976.
26. The wage increases were sizeable, a minimum of 8% for skilled workers. For more, see *Figyelö*, nos. 4 and 5, 1973.
27. *Müszaki Élet*, 7 June 1974.
28. I. Kertész, *Közgazdasági Szemle*, no. 9, 1977.
29. In this connection it is worthwhile mentioning D. Granick's contention that the goal of the reform to depersonalise control was frustrated by the various tax exemptions and subsidies. According to him, what really happened in Hungary was a shift in decision-making from branch ministries to interministerial committees which make decisions about enterprises' requests for subsidies and tax exemptions. See Granick, 1975, p. 310.
30. See F. Kölgyesi and A. Timár, *Pénzügyi Szemle*, no. 11, 1974.
31. O. Gadó, 1976, p. 60.
32. See also G. László and J. Veress, *Pénzügyi Szemle*, no. 3, 1978.
33. O. Gadó, 1976, pp. 54–5 and A. Timár, *Munkaügyi Szemle*, no. 12, 1975.
34. A. Timár, ibid.
35. A. Timár, *Munkaügyi Szemle*, nos. 1–2, 1978.
36. T. Sawczuk (*Gospodarka planowa*, no. 3, 1977) maintains that in 1976, 60% of enterprises applied this system.
37. For more, see Announcement no. 14, 1975, in *A gazdasági . . .*, pp. 67–9, also A. Timár, *Munkaügyi Szemle*, no. 12, 1975 and O. Gadó, 1976, p. 54.
38. A. Timár, *Munkaügyi Szemle*, no. 12, 1975, and G. Rák and A. Timár, *Munkaügyi Szemle*, no. 1, 1974.
39. See the formula for the division of profit on p. 152.
40. Produced profit is taxed by a single tax at 36%. The part of profit earmarked for the sharing fund is taxed in the following way: for funds corresponding to 2% of the wage-bill no tax is levied. For funds above 2% but not more than 4% of the wage-bill a tax of 200% is levied on the

additional funds. If the sharing fund reaches a level of 4–6% of the wage-bill, the cumulative tax will be 200% for the funds of 2–4% and 300% for the funds above 4%. With a sharing fund above 14% of the wage-bill an enterprise is obliged to pay 800% in taxes for the funds exceeding 14%. (*A gazdasági* ..., p. 197.) E.g., if an enterprise has, say, a wage-bill of 1000 money units and wants to have a sharing fund of 8% of the wage-bill (80 units), which is a modest size, then it has to pay 180 units in taxes.

41. *A gazdasági* ..., p. 197.
42. The tax on wage increases above what enterprises are entitled to on the basis of their performance ranges from 150–600% of the additional wage costs. The tax payable depends on the 'zone' in which the enterprise belongs due to its performance; e.g., if an enterprise is entitled on the basis of performance to an increase of 3% and it would like to have wages 1% higher, it will have to pay 350% of the additional wage costs. If an enterprise achieves a performance which, together with the guaranteed wage increase, entitles it to an increase above 6% (say 6.5%) it pays a tax of 150% on the difference. However, if it decides to increase wages above 6.5%, it must pay a progressive tax on the balance; in this case 600%. See O. Gadó, 1976, p. 52, and *A gazdasági* ..., pp. 64–5.
43. P. Bánki (see *Ipargazdaság*, no. 1, 1974) contends that the great labour reserves are not only due to the system of wage regulation.
44. See also L. Pongrácz, *Társadalmi Szemle*, no. 4, 1973.
45. A. Timár, *Munkaügyi Szemle*, nos. 1–2, 1978.
46. T. Sawczuk (*Gospodarka planowa*, no. 3, 1977) maintains that wage-bill regulation was applied only in 12% of enterprises.
47. A. Timár, *Munkaügyi Szemle*, no. 12, 1975. Profit is here defined in the same way as in the indicator of gross income per employee.
48. O. Gadó, 1976, p. 51.
49. For increases in the wage-bill above the three following items taken together, guaranteed increase, performance and labour saving, as long as they are smaller than 6%, a progressive tax in the range of 100–300% is payable. If an enterprise is entitled to an increase of 6.5%, it will pay for the $\frac{1}{2}$% only 100% of the additional wage costs. However, if it desires to increase wages above $6\frac{1}{2}$%, it is obliged to pay a progressive tax ranging from 100–300%. See O. Gadó, pp. 52–3 and *A gazdasági* ..., pp. 65–6.
50. *A gazdasági* ..., p. 68.
51. O. Gadó, 1976, pp. 55–6.
52. The tax rates are the same as in the case of regulation of average wages depending on performance.
53. A. Timár, *Munkaügyi Szemle*, no. 12, 1975.
54. Sz. Csaba, *Közgazdasági Szemle*, no. 3, 1977.
55. I. Kertész, *Közgazdasági Szemle*, no. 9, 1977, and L. Nyikos, *Közgazdasági Szemle*, no. 9, 1977.
56. L. Kónya, *Pénzügyi Szemle*, no. 2, 1978.
57. A. Timár, *Munkaügyi Szemle*, nos. 1–2, 1978.
58. L. Kónya, *Pénzügyi Szemle*, no. 2, 1978.
59. However, real wages in 1976 lagged behind the plan target (0.5% against 1.5%). It should, however, be mentioned that the target for national

income was not achieved either (3% against the planned 5–5.5%). See *Economic Survey of Europe for 1976*, part I, pp. 76 and 112.
60. A. Timár, *Munkaügyi Szemle*, nos. 1–2, 1978.
61. See L. Nyikos, *Közgazdasági Szemle*, no. 9, 1977.
62. A. Timár (*Munkaügyi Szemle*, nos. 1–2, 1967) indicates that it was expected that gross income per employee in industry would increase by 12%; in fact it increased by 17.5%.
63. Ibid.
64. Regulation methods:
    Application of different methods of wage regulation by number of employees (in %).

	1968	1971	1976	1978
Average wage	93	82	34	15
Wage-bill	7	9	35	55
Direct regulation of average wages	—	3	21	15
Direct regulation of wage-bill	—	6	10	15

SOURCE
J. Lökkös, *Közgazdasági Szemle*, no. 2, 1978

65. According to J. Lökkös, in 1976 employment in enterprises with wage-bill regulation declined compared with 1975, whereas in other enterprises it increased.
66. A. Timár, *Munkaügyi Szemle*, nos. 1–2, 1978 and L. Kónya, *Pénzügyi Szemle*, no. 2, 1978. J. Lökkös does not agree with such an evaluation. To him wage-bill regulation was the important factor for a better economy of labour.
67. J. Lökkös (*Közgazdasági Szemle*, no. 2, 1978) argues that enterprises with average wage regulation increase employment whenever possible. He does not explain this behaviour of enterprises. Does he have in mind that enterprises hire unskilled or less skilled workers in order to be able to increase the wages of skilled workers already on the payroll? Such behaviour is understandable if the wage growth indicator is not a productivity indicator. However, in Hungary average wage regulation is linked to gross income per employee.
68. A. Timár, *Munkaügyi Szemle*, nos. 1–2, 1978. This happened in enterprises which employ 15% of the labour force.

NOTES TO CHAPTER 10

1. See A. Bajcura, 1969, p. 38.
2. Ibid., pp. 40–6, and M. Sokol, *Plánované hospodářství*, no. 11, 1959.
3. See Z. Vergner, *Plánované hospodářství*, no. 3, 1957.
4. M. Sokol, *Plánované hospodářství*, no. 11, 1959.

5. A. Bajcura, 1969, p. 44, and M. Kaser and J. Zielinski, 1970, p. 122.
6. The expectations that enterprises might strive for an over-fulfilment of the productivity target (since this was rewarded by a higher normative) did not materialise. On the other hand, the target for industrial output was over-fulfilled by 4.4% due to increased employment. It turned out that enterprises preferred over-fulfilment of output over productivity, which is more difficult to achieve, and in addition reduces the wage-bill. For more see M. Sokol, *Plánované hospodářství*, no. 11, 1961.
7. All groups in an enterprise work-force cannot be equally responsible for economic returns. The responsiblity must be differentiated according to the relative importance and influence of different functional groups in the hierarchical structure of management. No doubt, managers must bear greater responsibility than blue-collar workers. Therefore the former must participate to a greater extent in both good and bad results.
8. See also M. Horálek, *Politická ekonomie*, no. 4, 1965.
9. The wage guarantee was set at approximately 92% of the previous year's total average wages.
10. Gross income was defined—to put it simply—as the sum which remains with the enterprise after the deduction of material costs, depreciation and the turnover tax from receipts resulting from sales of goods and services. In addition gross income usually included changes in inventories and in the value of unfinished products.
11. In this section wherever the term wage-bill and average wages are used, the first is to be understood as the basic wage-bill including all bonuses, whereas the second as an average resulting from the basic wage-bill including bonuses.
12. In January 1967 a uniform tax on gross income was introduced; for industry and construction it amounted to 18% and for domestic trade 32%; (it was later reduced). In addition, a further 6% tax was introduced on fixed assets, and a 2% tax on the value of inventories. (About the tax on wages, see below.) From gross income, enterprises also had to cover insurance expenses, penalties, payments of interest and contributions to the general management of the respective branch. The remainder was called 'resources of the enterprise'. From these resources, the respective enterprises had to finance the construction fund (which served to finance investment), the reserve fund (which served to replenish the employee fund in case the portion from gross income was not enough to cover wage claims), as well as the fund for cultural and social needs. Finally the balance was the wage-bill (the employee fund). (See *Sbírka zákonů*, no. 100, 1966).
13. See Z. Kodet, *Politická ekonomie*, no. 4, 1965. The tax ·on profit in industry amounted to 32%.
14. See F. Botek *et al.*, 1966, p. 33.
15. M. Sokol, *Plánované hospodářství*, nos. 8–9, 1966.
16. See also p. 94.
17. The tax in 1968 was also calculated in relation to 1966, but its real size was minus the tax payment for 1967.
18. In simple terms the formula meant that an enterprise which did not increase its average wages in relation to the planned wage for 1966 had to

pay three hellers for every one crown paid out in annual wages in 1967. In the case of average wage increases (the first division of the stabilisation tax) it had to pay an additional 30 hellers for every crown paid out additionally in wages. If the number of employees expanded (second division), the increase in the stabilisation tax was equal to the sum paid out in wages to new employees.
19. For more, see J. Adam, in *Jahrbuch der Wirtschaft Osteuropas*, vol. 3, 1972. The aim of the differentiation of tax rates was to hamper expansion of employment in branches with excessive manpower and thus to give sectors short of labour a greater share in the net addition to the labour force.
20. *Sbírka zákonů*, no. 63, 1966.
21. Z. Kutálek, *Plánované hospodářství*, nos. 8–9, 1966, and *Zásady urychlené realizace nové soustavy řízení*, 1966.
22. J. Typolt, *Plánované hospodářství*, no. 2, 1968.
23. For example, the average increase of prices in the mining industry, where average wages were the highest, amounted to 50%, whereas in the consumer goods industry which belonged to low wage industries, prices rose by only 20%. (See *Statistické přehledy*, no. 3, 1968.)
24. Based on information obtained from a big enterprise in Prague.
25. For more, see J. Adam, 1974, pp. 160–7.
26. 'The Government and the Trade Unions agreed that in 1969 the administrative and directive elements for regulating wage development in the *khozraschet* sphere would be eliminated, and that the binding relationship between the growth of productivity and the growth of wages for the purpose of controlling wage development would not be used.' This is a passage from the agreement between the Government and the presidium of the Trade Unions concluded on 11 November 1968. See *Práce a mzda*, no. 11, 1968. It is worthwhile mentioning that this happened approximately 3 months after the occupation of Czechoslovakia by the Soviet Union. This is no doubt good evidence of the Party's and the Government's determination to continue with the reforms despite the Soviet invasion.
27. Besides this surtax, in October 1968 the government had already introduced a new tax on profits at the rate of 45%. The purpose of this tax on profits was to skim off a part of the profits which enterprises had gained from the wholesale price reform.
28. A. Kudrna, *Práce a mzda*, no. 10, 1968.
29. The wage-bill is also used for payments of bonuses apart from the bonuses whch are paid from the bonus fund.
30. The bonus fund is used mainly for year-end rewards and for bonuses for the higher echelon of managers.
31. In Czech the indicator is called generally *výkony* (it means roughly output) and for industry *realizované výkony* (realised output). Because of its contents—value of output in realised prices—we term it marketed output. In most cases the term marketed output includes the value of unfinished products and inventories as well. Furthermore, it includes also the value of subcontracting services performed for other enterprises, but the turnover tax is not included. See J. Jílek *et al.*, 1976, p. 59.

32. See also M. Dąbrowski, *Gospodarka planowa*, no. 2, 1977.
33. See *Sbírka zákonů*, Announcement no. 158, 1970, in the working of no. 165, 1970, as published in *Práce a mzda*, no. 3, 1972.
34. See *Sbírka zákonů*, Announcement no. 157, 1975, as published in *Práce a mzda*, no. 3, 1976.
35. Ibid.
36. See also E. Moravec, *Práce a mzda*, no. 3, 1974.
37. J. Hejkal, *Politická ekonomie*, no. 9, 1974.
38. See Supplement to *Hospodářské noviny*, no. 50, 1975, and T. Hauptvogel, *Práce a mzda*, no. 6, 1976.
39. T. Hauptvogel, *Práce a mzda*, no. 7, 1974.
40. L. Lér, *Plánované hospodářství*, no. 1, 1978, and also *Hospodářské noviny*, no. 2, 1978.
41. Ibid. and also S. Šourek, *Finance a úvěr*, no. 3, 1978.
42. In Czech *vlastní výkony*. For the time being a detailed definition of net output is not available. What the government resolution stresses is that the computation of net output should not be affected by foreign trade. See Supplement to *Hospodářské noviny*, no. 2, 1978.
43. A second minor difference is that the method of two adjustment coefficients was dropped and again only one exists.
44. See, e.g., E. Moravec, *Práce a mzda*, no. 31, 1974; J. Dvořák, *Politická ekonomie*, no. 4, 1977; I. Címa, *Plánované hospodářství*, no. 5, 1976.
45. Supplement to *Hospodářské noviny*, no. 2, 1978.
46. See S. Šourek, *Finance a úvěr*, no. 3, 1978. Šourek's views seem to differ from the Government views as to the purpose of the new bonus fund.
47. Part of the funds will be derived from savings achieved in the basic wage-bill, from profit made on higher prices allowed for higher quality goods and from the fund of material interest in export (a new fund for the purpose of promoting export has been established).
48. *Sbírka zákonů*, Announcement no. 157, 1975.

NOTES TO CHAPTER 11

1. See P. Wiles, *Economie appliquée*, no. 1, 1976.
2. See also Table 1.4. In Hungary, as already mentioned, of the 18% by which consumer prices increased in 1968–75, 10% was due to changes in free prices.
3. P. Wiles, 1974, p. 240.
4. Ibid.
5. Figures in Table 2.5 are macroeconomic figures.
6. In a book which appeared in Prague, J. Valach (1972, p. 33) characterizes marketed output in the following way: '... it orients the material interest of enterprises to an increase of output as an end in itself regardless of the needs of the buyers, economic efficiency and social labour productivity.'
7. J. Valach, (ibid.) maintains that 'Shares in the economic returns which link the level of bonuses with the profit of enterprise cannot essentially change this fundamental tendency" (he is referring to the quotation indicated in Note 6).

8. For more, see R. Portes, 1977, pp. 771–3.
9. See *Figyelö*, no. 21, 1978, p. 1.
10. K. Golinowski, in *Nowy system ekonomiczno-finansowy* . . . , p. 33, and K. Porwit, in *Zarys systemu funkcjonowania* . . . , p. 62.
11. This statement is true only generally as long as it refers to the wage-bill including the bonus fund. But there may be quite a conflict of interest between managers and workers—as the Hungarian system of 1968–9 showed—as to the distribution of the funds between basic wage-bill and bonuses.
12. B. Fick, 1965, p. 313.
13. Ibid., p. 312; see also J. Zielinski, 1973, p. 260.
14. See O. Šik, 1973, p. 101.
15. V. Měchura, *Práce a mzda*, no. 7, 1973, and Z. Jacukovicz, 1974, p. 96.
16. Up to the 1960s, particularly in the USSR and Poland, progressive piece-rates were relatively widely used. In Poland they were an important contributing factor to inflation.
17. There is a tendency in all the countries to reduce the role of piece-rates, which is the largest in the USSR. According to Shkurko (1975, p. 103), the share of piece-rate workers in industry is expected to decline in 1976–80 to 40–50%.
18. See *Sbírka zákonů, ČSSR*, no. 157, 1975, published in *Práce a mzda*, no. 2, 1976.
19. According to V. Rakoti (*Sotsialisticheskii trud*, no. 12, 1977) inspections carried out in 1976 revealed that up to 15% of individual types of light industry goods were rejects.
20. J. Lökkös, *Közgazdasági Szemle*, no. 2, 1978.
21. G. Grossman, 1977, p. 140.
22. P. Marton, *Gazdaság*, no. 3, 1977. It is not clear how P. Marton arrived at the figure 19%. What is clear is that the normative is small.
23. See also D. Granick, 1975, pp. 309–16.
24. See, e.g., B. Gliński, *Gospodarka planowa*, no. 7, 1974.

NOTES ON CONCLUSION

1. Some Hungarian economists are disappointed with the working of the wage regulation system. They hoped it would induce enterprises to use labour more rationally. In the present system the linkage of wage growth to performance over the previous year ensures enterprises funds for wages even if their efficiency is below average with the result that they are not under pressure to economise on labour. Confronted with labour shortages central authorities have resorted to some direct methods in the allocation of labour which is in contradiction with the principles of the reform of 1968. Therefore some authors suggest embarking on a direct assignment of funds for wages to enterprises from the centre and using it as an indirect instrument for allocation of labour according to the efficiency of enterprises. See for example D. Bonifert, *Figyelö*, no. 4, 1977.
2. This problem is especially acute in Hungary due to growing labour shortages.

Appendix

I CURRENT REGULATION OF WAGES IN THE SOVIET BLOC COUNTRIES

	Objects of regulation	Formation of the basic wage-bill		Regulation of wage-bill or wages
		Planned	Actual-depends on	
USSR	wage-bill	assigned as an absolute sum	fulfilment of plan targets in gross value of output	over-fulfilment (under-fulfilment) of targets increases (decreases) the wage-bill by an adjustment coefficient
GDR	wage-bill	assigned as an absolute sum	fulfilment of plan targets in commodity production per employee	"
Czechoslovakia	wage-bill	assigned as a planned proportion of marketed output	fulfilment of plan targets in marketed output	"
Poland[a]	wage-bill	centre assigns only indicators and normatives	increase in output added over the previous year and normative	(1) determining the indicator and normative (2) by charges on the disposable wage-bill
Hungary[b]	(a) wage-bill		(a) increase in value added over the previous year and normative	(1) determining the indicator and normative
	(b) average wages	centre assigns only indicators and normatives	(b) increase in gross income per employee over the previous year and normative	(2) by taxes payable from the bonus fund

[a] Modifications of 1976–7 are considered

II MODELS OF WAGE REGULATION IN THE SOVIET BLOC COUNTRIES

	Objects of regulation	Formation of the basic wage-bill		Regulation of wages
		Planned	Actual-depends on	
direct	wage-bill	assigned as an absolute sum or as proportion of plan target	fulfilment of the targets	over-fulfilment (under-fulfilment) of targets increases (decreases) the wage-bill by an adjustment coefficient
mixed	wage-bill	centre assigns only indicator(s) and normative(s)	increase in performance over the previous year in terms of the indicator	by determining the indicator and normative
indirect[a]	can be both, average wages have some advantages	centre determines the indicator; enterprises determine the size of the wage-bill	performance	by taxes

[a] In the indirect system enterprises themselves determine the distribution of the wage-bill into basic wages and bonuses.

Selected Bibliography

(Articles published in other languages than English and National Statistical Yearbooks are not listed.)

Adam, J., *Wage, Price and Taxation Policy in Czechoslovakia 1948–1970*, Berlin, 1974(a).
——, 'The Incentive System in the USSR', *Industrial and Labour Relations Review*, October 1973.
——, 'The System of Wage Regulation in Hungary', *Canadian Journal of Economics*, November 1974(b).
——, 'The Recent Reform of the Incentive System in Poland', *Osteuropa Wirtschaft*, no. 3, 1975.
A gazdasági szabályozó rendszerrel kapcsolatos jogszabályok és magyarázatok, Budapest, 1975.
Ames, E., 'Soviet Bloc Currency Conversions', *American Economic Review*, June 1954.
Bácskai, T., 'New Development in State Enterprise Taxation in Hungary'. *Acta Oeconomica*, vol. 6, nos. 1–2, 1971.
Bajcura, A., *Hmotná zainteresovanost v priemysle*, Bratislava, 1969.
Balassa, B., 'The Firm in the New Economic Mechanism in Hungary', in *Plan and Market*, Bornstein, M. (ed.), Yale, 1973.
Batyrev, V. M. (ed.), *Kreditnoe i kassovoe planirovanie*, Moscow, 1947.
Baykov, A., *The Development of the Soviet Economic System*, Cambridge, 1947.
Bér- és jövedelemarányok az iparban, 1967–71, Hungarian Statistical Board, 1973.
Berend, T. I., *A szocialista gazdaság fejlödése Magyarországon 1945–1968*, Budapest, 1974.
Berényi, J., *Lohnsystem und Lohnstruktur in Österreich und in Ungarn*, Vienna, 1974.
Berliner, J. S., *Factory and Manager in the USSR*, Cambridge, USA, 1957.
——, *The Innovation Decision in Soviet Industry*, MIT, 1976.

Berri, L. (ed.), *Planning of a Socialist Economy* (translation from Russian), Moscow, 1973.
Bornstein, M., 'Soviet Price Policy in the 1970's', in *Soviet Economy in a New Perspective*, US Congress, 1976.
Botek, F., Pokorný, J. and Voják, J., *Odbory a hmotná zainteresovanosť v novej sústave riadenia*, Bratislava, 1966.
Brus, W., *The Market in a Socialist Economy*, London, 1972.
Buda, I. and Pongrácz., L., *Személvi jövedelmek, anyagi érdekeltség, munkaerö-gazdálkodás*, Budapest, 1968.
Bury, J., Karwanski, R. and Migdal, T., *Place i premie*, Warsaw, 1976.
Butakov, D. D., *Finansove problemy khoziaistvennykh reform v stranakh chlenakh SEV*, Moscow, 1973.
Campbell, R., 'Economic Reform in the USSR', *American Economic Review*, May 1968.
Chapman, J. G., *Real Wages in Soviet Russia Since 1928*, Harvard, 1963.
Cohn, S., 'National Income Growth Statistics', in *Soviet Economic Statistics*, Treml, B. G. and Hardt, H. P. (eds.), Duke, 1972.
Csikós-Nagy, B., *Magyar gazdaságpolitika*, Budapest, 1971.
——, 'New Aspects of the Profit Incentive', *The New Hungarian Quarterly*, Winter 1965.
——, 'Profit in a Socialist Economy', *Economie appliquée*, no. 4, 1972.
——, 'Anti-Inflationary Policies Debates and Experience in Hungary', *CESES*, L'Est, no. 6, 1974.
DDR—Wirtschaft, Frankfurt, 1974.
Dobb, M., *Soviet Economic Development since 1917*, London, 1966.
Egiazarian, G. A., *Materialnoye stimulirovanie rosta effektivnosti promyshlennogo proizvodstva*, Moscow, 1976.
Ellman, M., *The Economic Reform in the Soviet Union*, PEP, 1969.
Fallenbuchl, Z., 'Commecon Integration', *Problems of Communism*, March–April, 1973(a).
——, 'The Strategy of Development and Gierek's Economic Maneuver', *Canadian Slavonic Papers*, nos. 1–2, 1973(b).
Farrell, J. P., 'Bank Control of the Wage-bill in Poland, 1950–70', *Soviet Studies*, April 1975.
Fedorenko, N. and Bunicha, P., *Mekhanizm ekonomicheskogo stimulirovaniia pri sotsializme*, Moscow, 1973.
Fedorowicz, Z., *Finanse organizacji gospodarczych*, Warsaw, 1977.
Ferge, S. and Antal, L., 'Enterprise Income Regulations', in *Reform of the Economic Mechanism in Hungary, Development 1968–71*, Gadó, O. (ed.), Budapest, 1972.
Fick, B., *Bodźce ekonomiczne w przemysle*, Warsaw, 1965.
——, *Fundusz zakladowy*, Warsaw, 1967.
Finansovo-kreditnii kontrol v narodnom khoziaistve, Riga, 1975.
Flakierski, H., 'The Polish Economic Reform of 1970', *Canadian Journal of Economics*, no. 1, 1973.
Flór, F. and Horváth, P., *Iparvállalatok jövedelem- és munkaerogazdálkodása*, Budapest, 1972.
Foglalkoztatottság és kereseti arányok, 1973, Hungarian Statistical Board, 1975.

Friss, I., 'Principal Features of the New System of Planning, Economic Control and Management in Hungary', in *Reform of the Economic Mechanism in Hungary*, Friss, I. (ed.), Budapest, 1969.

Gadó, O., *Közgazdasági szabályozó rendszerünk 1976-ban*, Budapest, 1976.

Garmarnikov, M., 'A New Economic Approach', *Problems of Communism*, Sept.–Oct. 1972.

Garvy, G., *Money, Banking and Credit in Eastern Europe*, New York, 1966.

——, 'Inflation and Price Stabilization Policies in some Eastern European Countries', CESES, L'Est, no. 6, 1974.

Gazdaságpolitikánk tapasztaltai és tanulságai 1957–1960, Friss, I. (ed.), Budapest, 1976.

Gliński, B., Kierczyński, T. and Topiński, A., *Zmiany w systemie zarzadzania przemyslem*, Warsaw, 1975.

Granick, D., 'Variations in Management of the Industrial Enterprise in Socialist Eastern Europe', in *Reorientation and Commercial Relations of the Economies of Eastern Europe*, US Congress, 1974.

——, *Enterprise Guidance in Eastern Europe*, Princeton, 1975.

Grossman, G., 'Price Control, Incentives and Innovation in the Soviet Economy', in *The Socialist Price Mechanism*, Abouchar, A. (ed.), Duke, 1977.

Gruzinov, V. P., *Materialnoe stimulirovaniie truda v stranakh sotsialisma*, Moscow, 1968.

Hensel, K., Wessely, K. and Wagner, U., *Das Profitprinzip—seine ordnungspolitischen Alternativen in sozialistischen Wirtschaftssystemen*, Stuttgart, 1972.

Hodgman, D. R., *Soviet Industrial Production 1929–51*, Harvard, 1954.

Holzman, F., 'Soviet Inflationary Pressures, 1928–57', *Quarterly Journal of Economics*, no. 2, 1960.

——, *Soviet Taxation*, Cambridge, USA, 1962.

Howard, D., 'A Note on Hidden Inflation in the Soviet Union', *Soviet Studies*, no. 4, 1976.

Incomes in Postwar Europe: A Study of Policies, Growth and Distribution, U.N. Geneva, 1967.

Jacukowicz, Z., *Proporcje plac w Polsce*, Warsaw, 1974.

Jasny, N., *The Soviet Price System*, Stanford, 1951.

——, *Soviet Industrialization, 1928–52*, Chicago, 1961.

Jaworski, W. and Sochacka-Krysiak, H., *Finanse krajów socjalistycznych*, Warsaw, 1977.

Jílek, J. et al., *Příručka statistiky pro hospodářské pracovníky*, Prague, 1976.

Kabaj, M., 'Evolution of the Incentive System in the USSR Industry', *International Labour Review*, no. 1, 1966.

Kaser, M. and Zielinski, J. G., *Planning in East Europe*, London, 1970.

Katsenelinboigen, A., 'Disguised Inflation in the Soviet Union', in *Economic Aspects of Life in the USSR*, NATO, 1975.

Khoziaistvennaia reforma v SSSR, Moscow, 1969.

Kierczyński, T. and Wojciechowska, U., *Rola zisku w systemie ekonomicznofinansowym*, Warsaw, 1972.

Kirsch, L. J., *Soviet Wages*, MIT, 1972.

Kohlmey, G. and Dewey, Ch., *Banksystem und Geldumlauf in der DDR, 1945–1955*, Berlin, 1957.

Kónya, L., 'Further Improvement of the System of Enterprise Income and Wage Regulation', *Acta Oeconomica*, no. 1, 1971.
Köszegi, L., 'Recent Price and Income Trends in Hungary', *International Labour Review*, no. 2, 1975.
Kotowitz, Y. and Portes, R., 'The Tax on Wage Increases', *Journal of Public Economics*, no. 3, 1974.
Közgazdasági Kislexikon, Budapest, 1968.
Kožušník, C., *Problémy teorie hodnoty a ceny za socialismu*, Prague, 1964.
Krencik, W., *Podstawy i kierunki polityki plac w PLR*, Warsaw, 1972.
——, *Place a wzrost gospodarczy*, Warsaw, 1977.
Kronrod, Ya., *Ukreplenie denezhnogo obrashcheniia*, Moscow, 1950.
Krueger, C., 'A Note of the Size of the Subsidies on Soviet Government Purchases of Agricultural Products', *ACES Bulletin*, no. 2, 1974.
Kucharski, M., *Pieniądz, dochod. proporcje wzrostu*, Warsaw, 1972.
Laski, K., *Nastin teorie socialistické reprodukce* (translation from Polish), Prague, 1967.
Levčik, B., *Wage Policy and Wage Planning in Czechoslovakia*, 1969, (unpublished paper).
Lexikon der Wirtschaft, Arbeit, Berlin, 1968.
Liberman, E., 'The Role of Profits in the Industrial Incentive System of the USSR', *International Labour Review*, no. 1, 1968.
Maier, V. F., *Zarabotnaia plata v period perekhoda k kommunizmu*, Moscow, 1963.
——, *Uroven zhizni naseleniya SSSR*, Moscow, 1977.
Marton, A., 'Trend of Consumer Prices in Hungary 1968–1975', *Acta Oeconomica*, v. 14, (4), 1975.
McAuley, M., *Labour Disputes in Soviet Russia 1957–1965*, Oxford, 1969.
Meller, J., *Place a planowanie gospodarcze w Polsce 1950–75*, Warsaw, 1977.
Miliukov, A. I., *Mekhanizm stimulirovaniia rosta effektivnosti truda*. Moscow, 1977.
Montias, J., 'Inflation and Growth: The Experience of Eastern Europe', in *Inflation and Growth in Latin America*, Baer, W. and Kerstenetzky, I. (eds.), Yale, 1964.
Nagy, T., *Az árak szerep a szocializamusban*, Budapest, 1960.
Nogaro, B., 'Hungary's Recent Monetary Crisis and its Theoretical Meaning', *American Economic Review*, September 1948.
Nove, A., *An Economic History of the USSR*, Pelican, 1972.
——, *The Soviet Economic System*, London, 1977.
Nowy system ekonomiczno-finansowy w organizacjach przemyslowych, Sliwa, J. (ed.), Warsaw, 1974.
Ökonomik der Arbeit, Sixth edition, East Berlin, 1974.
Oplata truda pri sotsializme: voprosy teorii i praktiki, Moscow, 1977.
Petrochenko, P. F. et al., *Ekonomika truda*, Moscow, 1965.
Planirovanie narodnogo khoziaistva SSSR, Tsapkin, N. V. and Pereslegin, V. I. (eds.), Moscow, 1967.
Popov, V. F. (ed.), *Gosudarstvenii Bank SSSR*, Moscow, 1957.
Portes, R. D., 'Economic Reform in Hungary', *American Economic Review*, May 1970.

——, 'Hungary: Economic Performance, Policy and Prospects', in *East-European Economies Post-Helsinki*, US Congress, 1977.
——, 'The Control of Inflation', *Economica*, May 1977.
Pruss, W., *Ceny*, Warsaw, 1975.
Řízení národního hospodářství, Bránik, J. (ed.), Prague, 1967.
Schroeder, G., 'Recent Developments in Soviet Planning and Incentives', in *Soviet Economic Prospects for the Seventies*, US Congress, 1973.
——, 'Consumer Goods Availability and Repressed Inflation in the Soviet Union', in *Economic Aspects of Life in the USSR*, NATO, 1975.
Schroeder, G. and Severin, B., 'Soviet Consumption and Income Policies in Perspective', in *Soviet Economy in New Perspective*, US Congress, 1976.
Shurko, S. I., *Sovershenstvovanie form i system zarabotnoi platy*, Moscow, 1975.
Sielunin, V., *Reforma gospodarcza w ZSRR* (translation from Russian), Warsaw, 1971.
Šik, O., *Argumente für den Dritten Weg*, Hamburg, 1973.
Sitnin, V., *Chistii dokhod*, Moscow, 1974.
Spulber, M., *The Economics of Eastern Europe*, N.Y., 1957.
Statisticheskii sbornik po voprosam truda i zarabotnoi platy v evropeyskikh sotsialisticheskikh stranakh, Moscow, 1959.
Stručný hospodářský vývoj Československa do roku 1955, Prague, 1969.
Struktur- und stabilitätspolitische Probleme in alternativen Wirtschaftssystemen, Watrin, Ch. (ed.), 1974.
Sulyok, B., 'Major Financial Regulators in the New System of Economic Control and Management', in *Reform of the Economic Mechanism in Hungary*, Friss, I. (ed.), Budapest, 1969.
System und Entwicklung der DDR-Wirtschaft, Deutsches Institut fur Wirtschaftsforschung, Berlin, 1974.
Teoretické základy a praxe hospodářské reformy v SSSR, (translation from Russian), Prague, 1973.
Timár, M., *Gazdaságpolitika Magyarországon 1967-73*, Budapest, 1973.
Tomášek, P., *Odměňování v nových podmínkách řízení*, Prague, 1967.
Trud i zarabotnaia plata v SSSR, Volkov, A. P. (ed.), Moscow, 1975.
Typolt, J., Janza, V. and Popelka, M., *Plánování cen průmyslových výrobků*, Prague, 1959.
Valach, J., *Finance a finanční rozhodování průmyslových podniků*, Prague, 1972.
Vincze, I., *Árak, adók, támogatások a gazdaságirányítás reformja után*, Budapest, 1971.
Wagner, U., See Hensel, K., Wessely, K. and Wagner, U.
Ward, B., 'The Firm in Illyria, Market Syndicalism', *American Economic Review*, no. 4, 1958.
Wiles, P., 'The Inflation consists of too many bankers chasing too few ideas', CESES, *L'Est*, no. 6, 1974.
——, *Economic Institutions Compared*, Oxford, 1977.
Wirtschaftssysteme des Soziaismus im Experiment—Plan oder Markt? Bress, L. and Hensel, K. P. (eds.), Frankfurt, 1972.
Zaleski, E., *Planning for Economic Growth in the Soviet Union, 1918-32*, 1971.
Zarys systemu funkcjonowania przemysłowych jednostek inicjujących, Gliński, B. (ed.), Warsaw, 1975.

Zásady uryhlené realizace nové soustavy řízení, Materials of the Committee for Economic Reforms, Prague, 1966.

Zielinski, J. G., *Economic Reforms in Polish Industry*, Oxford, 1973.

Index

The letter (d) stands for 'defined'

Abramov, G., 216
Adam, J., 197–8, 210–12, 215, 219, 222, 226
Adamíček, J., 22, 197
Adjustment coefficient, 68, 125–6, 133, 135, 138, 164, 171, 173, 187
Ames, E., 12, 198
Antal, L., 222
Artemov, Iu., 89, 210–11

Bácskai, T., 212, 221
Bajcura, A., 224–5
Bakhurin, A., 96, 212
Balassa, B., 222
Baldwin, G., 202
Balousek, A., 211
Bank control, 58, 72–4
  in Czechoslovakia, 73, 174
  in the GDR, 136
  in Hungary, 73
  in Poland, 149–50
  in the USSR, 73, 123, 132–4, 187
Bánki, P., 160, 209, 221–3
Barkovskii, N., 216
Barngolc, S., 216
Batkaev, R., 205
Batyrev, V., 216
Baykov, A., 198

Berend, T. I., 198, 204–5
Berényi, J., 207, 215
Berliner, J. S., 206, 208, 211
Berri, L., 197
Beskid, L., 200
Blass, B., 217–18
Bokor, J., 222
Bonifert, D., 228
Bonus fund regulation, XV, 58, 77–85, 104, 111
  direct, 80
  in Czechoslovakia, 69–70, 82–3, 85, 165, 167, 171, 173, 183
  in the GDR, 78, 80–1, 84, 135, 183
  in Hungary, 79, 85, 152–4, 157, 162
  in Poland, 84–5, 140, 143
  in the USSR, 78, 80–1, 84, 127–30, 183
  indirect, 79–80, 84
  mixed, 78, 80, 84
Bonuses
  a part of average wages, 88–91
  and inflation, 88, 91
  for top managers, 88–91, 153
Bornstein, M., 18, 200
Botek, F., 225
Brezhnev, L., 19
Bronson, D., 45
Brus, W., 196
Buda, I., 213, 221
Bukhanevich, B., 205

Index

Bunich, P., 209, 213
Bunich, P. G., 89
Bury, J., 207, 213, 219, 221
Butakov, D., 213

Campbell, R., 215
Cernes, A., 216
Chapman, J., 197
Charges on wages
  compared to taxes, 72, 115, 149
  in Poland, 72, 115, 148, 191–2
Cichowski, E., 219
Címa, I., 211, 227
Clarke, R., 18
Cohn, S., 203
Collective agreements, 58, 74–6
Commodity production as indicator, XX, 80–1, 102, 105, 125, 135, 137, 209(d)
Control by rouble, 73
Cost of living index, 18–19, 30
Csaba, Sz., 223
Csikós-Nagy, B., 97, 99, 102, 199, 201, 212, 214, 220
Currency reforms, 4, 9–14

Dąbrowski, M., 227
Derco, M., 203
Dewey, Ch., 217
Dobb, M., 7, 197–8
Dubček, A., 30, 105, 170, 190
Dvořák, J., 216, 227

Egiazarian, G., 119, 129, 214–15
Ellman, M., 215
Employment
  control, 52, 65, 112, 148, 156, 168, 186
  growth and plan targets, 46, 48–50, 52, 178, 190
Ernst, P., 212
Érsek, T., 200

Factory fund, 89, 138, 142
Fallenbuchl, Z., 196, 201
Farrell, J. P., 218
Fedorenko, N., 89
Fedorowicz, Z., 219
Ferge, S., 222
Fick, B., 185, 189, 206–7, 212–13, 217–19, 228

Fiszel, H., 212
Flakierski, H., 218
Flór, F., 221–2
Fogl, J., 202
Friss, I., 221

Gadó, O., 222–3
Gamarnikov, M., 196
Garetovskii, N., 214
Garvy, G., XVI, 6, 196, 208
Giebner, J., 217
Gierek, E., 28, 141
Gliński, B., 219–20, 228
Glowacki, R., 200
Glushkov, N., 19
Golinowski, K., 150, 214, 219–21, 228
Gomulka, W., 22, 28
Gorski, R. N., 201
Granick, D., 86–8, 210–11, 214, 222, 228
Gross income as indicator, 69, 93, 95–100, 102, 144, 166–7, 225(d)
Gross income per employee as indicator, 154, 158–9, *see also* Productivity as indicator
Gross value of output as indicator, XX, 71, 80, 102, 105, 125–6, 130–1, 133–4, 137, 171–2, 182–3, 208(d)
Grossman, G., 197, 199–200, 228
Gruzinov, V., 207

Haffner, F., 200
Hauptvogel, T., 227
Havránek, V., 206
Hejduk, S., 202
Hejkal, J., 227
Hensel, K. P., 135, 210, 217
Hodgman, D. R., 197
Holubicki, B., 220
Holzman, F., 197, 214, 216
Horálek, M., 225
Hornok, B., 196
Horváth, P., 221–2
Howard, D., 19, 200

Ignatushkin, V., 209
Incentive system for top managers, 85–8
Inflation, XIV–XVIII, 5–8, 15, 17–19, 26, 31, 33–4, 37, 44, 46, 48, 51–2, 64–7, 76–7, 103, 114, 178, 181–2, 186, 192, 195

and systems of management, 179–81
  in Czechoslovakia, 3, 6, 11, 17, 25, 30, 39, 42, 168–9, 190
  in the GDR, 3, 11, 17, 20, 37, 190
  in Hungary, 3, 6, 25, 39–40, 43, 151, 158
  in Poland, 3, 6, 11, 25, 30–2, 39–41, 43, 137, 145–6, 228
  in the USSR, 3, 6, 17, 19–20, 37, 39, 127, 132, 190
  wage inflation, 33(d), 34–5, 58, 67, 70, 156, 178, 190, 192
Investment overstrain, *see* Overinvestment

Jacukowicz, Z., 211, 228
Jagielski, M., 28
Jakubovicz, S., 218
Janza, V., 198
Jasny, N., 7, 197–8
Jaworski, W., 213
Jędrychowski, S., 220
Jílek, J., 197, 209, 226

Kabaj, M., 206, 214, 219
Kádár, J., 22
Kalisiak, J., 214
Kamiński, M., 220
Kapustin, E., 130, 207, 216
Karpukhin, D., 107, 213
Kaser, M., 225
Katsenelinboigen, A., XVI, 196, 199
Kemény, A., 222
Kertész, I., 222–3
Khromov, P., 216
Khrushchev, N., 19, 21
Kierczyński, T., 207, 212–13
Kiperman, G., 216
Kirsch, L. J., 215
Kiss, O., 206
Kletskii, V., 215
Klocek, K., 219
Kodet, Z., 95–6, 212, 225
Kohlmey, G., 217
Kölgyesi, F., 213, 222
Kónya, L., 222–4
Kos, Cz., 200
Kosta, J., 95, 212
Köszegi, L., 201
Kotlaba, M., 203

Kotowitz, Y., 214
Kožušník, Č., 6, 198
Krajewski, B., 218
Krajkovič, T., 217
Krencik, W., 202–3, 205, 207
Krizsanits, J., 209
Kronrod, Ia., 198
Krueger, C., 200
Kucharski, M., 198–9, 202, 204–5
Kudrna, A., 226
Kunelskii, L., 216
Kutálek, Z., 226

Labour
  hoarding, 39, 51, 99, 112, 154–6, 185–6
  plan, 124
  rationing, 49, 51
Ladányi, J., 200
Langer, H., 20
Laski, K., 202
László, G., 222
Lér, L., 227
Levčik, B., 95, 203, 212
Liberman, E., 96, 99, 101, 212, 215
Lökkös, J., 66–7, 207–8, 224, 228
Lukeš, V., 204

Maier, E., 98
Maier, V., 125, 205, 207, 215
Majtan, M., 210
Managerial system, 96, 100
Margulis, Iu., 125–6, 210–11, 214–16
Marketed output as indicator, 105, 171, 173, 182, 226(d)
Marton, A., 27, 190, 200–1
Marton, P., 228
Marzec, J., 208
Maximising model, 86–87
Měchura, V., 228
Meitner, T., 221
Meller, J., 46–7, 50, 202, 205–6, 218
Melzer, M., 200
Mieszczankowski, M., 220
Miliukov, A., 80, 209, 211, 214
Montias, J., 198, 205
Moravec, E., 210–11, 213, 227
Moskalenko, V., 215
Mujżel, J., 200
McAuley, M., 208

Index

Nagy, T., 197
National income accounting, Marxist concept, XX, 36, 203
Nedzowski, E., 219
Nemchinov, V., 99, 212
Net output as indicator, 98, 102, 105, 130–1, 173
Nick, H., 98, 212
Nogaro, B., 198
Normatives
  defined, 67–8
  long-term, XV, 116–19
    in Czechoslovakia, 117, 165
    in the GDR, 83, 117, 135
    in Hungary, 117–19
    in Poland, 117–19, 144, 149
    in the USSR, 83–4, 117–19, 128–9
Nove, A., 199, 213
Nyikos, L., 223–4

Obodowski, J., 219
Olszewski, J., 208
Opaliński, K., 219
Output added as indicator, 95, 100, 107, 141, 142(d), 145–8
Over-investment, XVI–XVIII, 49
  and inflation, XVI–XVII, 30, 51
  effect on employment, 52
  other consequences, XVII

Pajestka, J., 95, 212
Pervukhin, A., 96, 212
Petrochenko, P., 215
Piece-rates, 40
  progressive, 40–1, 61, 126, 205(d), 228
  regressive, 187–8
  straight, 186–7
Pithe, E., 198
Plan targets, over-fulfilment and inflation, 127
  other consequences, 70, 126–7
  unsaleable goods, 127, 138, 187
Pohl, H., 98
Pongrácz, L., 213, 221–3
Popelka, M., 198
Popov, V., 133, 214, 216
Portes, R., 45, 75, 200, 208, 214, 228
Porwit, K., 228
Price(s)
  agricultural, 19, 24, 26–8
  commercial, 8–9, 11–12

  cutting, 3–4, 9, 14–16, 21–2, 37
  distortions, XIV, 6, 19, 25, 101, 180
  index, 19
  manipulation, 146, 189
  policies, XIV, XVI, 3–4, 17–30, 43, 180, 192
  reforms, 4, 10, 20, 25, 29, 129, 151, 169–70
  relativities, 23–7, 31–2, 101
  Soviet-type, 4–6
  subsidies, 5, 19–20, 24, 27–8
Productivity as indicator, 81, 99, 106–7, 135–6, 147, 163–4, 169–70, 173
Profit as indicator, XV, XX, 79, 81, 93–102, 104–5, 128–9, 140, 143, 157, 165–7, 171
Profitability as indicator, 99–102, 105, 128–9, 140
Pruss, W., 201

Rák, G., 222
Rakoti, V., 228
Risina, G., 215
Rogovskii, N., 216
Roubal, L., 212
Rozhkova, V., 107, 213
Rudcenko, S., 45
Rybnikář, K., 204
Rzheshevskii, V., 209–10, 214–15

Safirova, V., 217
Sales as indicator, XX, 84, 102–3, 128–9, 171
Satisficing model, 86–7
Sawczuk, T., 214, 222–3
Schönherz, H., 217
Schroeder, G., 19, 45, 196–7, 200, 209, 211, 215
Schuster, I., 217
Schwarz, B., 196, 202
Seifert, E., 98, 212
Sekanina, V., 217
Self-management system, 96, 99–100
Severin, B., 19, 45, 200
Shchekino experiment, 132, 207
Shkurko, S., 206–7, 228
Sielunin, W., 215
Šik, O., 196, 228
Sitnin, V., 98, 105, 212–16
Slowecki, H., 219
Smith, A. J., 201

Index

Sochacka-Krysiak, H., 213
Sokol, M., 94–6, 211, 224–5
Šourek, S., 212, 227
Spulber, N., 13, 198
Stalin, J., 10, 21, 40
Strategy of growth, XVII
  extensive, 39, 66
  intensive, 66
Struminski, J., 23, 201
Success indicators
  gross, 102–4, 188, 193
  net, 102–4, 106, 193, 212–13(d)
  possible manipulation, 103, 146, 188
  their grouping, XIX–XX
  their role in different systems, XIX, 92–3
Suchá, A., 210–11, 217
Sucharevskii, B., 205–7, 214, 216
Sulyok, B., 97, 212, 221
Szávai, A., 207, 210
Sztyber, W., 197, 200

Taxation
  criterion of efficiency, 94, 114
  in different systems, 112, 194
  instrument of bonus fund control, 85, 143, 152, 157
  instrument of wage control, 69, 71–3 111–15
    in Czechoslovakia, 70, 73, 85, 114–15, 167–70, 190
    in Hungary, 70, 73, 114–15, 155–7, 159–62, 190–1
Timár, T., 158–9, 212–14, 222–4
Toms, M., 202
Topiński, A., 208–9, 211, 219–20
Trade Unions, 74–5, 179–80
Turnover tax, 4–5, 21, 23, 26, 29, 145–6
Two-channel system, 78–9
Typolt, J., 198, 226

Valach, J., 213, 227
Value added as indicator, 102, 160–1(d)
Veress, J., 222
Vergner, Z., 224
Vincze, I., 21
Vlach, P., 199, 202

Wage
  average, defined, XIX
  basic average, defined, XIX

maximisation, 103, 138, 185–6, 188
plan over-expenditure, 52, 72–3, 133–4, 136, 138, 149, 174, 178, 182
  due to average wage growth, 52, 178
  due to employment growth, 52, 178, 190
Wage-bill
  basic, defined, XVIII–XIX
  disposable wage-bill in Poland, 142
  enterprise, defined, XVIII
  integrated, 78, 92
Wage control, XIV, 51, 57–8(d), 59–76, 102–7, 111, 177, 181, 185, 196
  and consumer demand, 182–5
  functions, their grouping, 58
  in Czechoslovakia, 68–9, 105, 164–74, 192
  in the GDR, 68, 105, 134–6, 192–3
  in Hungary, 69–70, 72, 106, 151–63, 192–4
  in Poland, 68, 72, 106, 137–50, 194
  in the USSR, 67–8, 105, 123–7, 130–4
  its effectiveness, 181–92
  of average wages, 58, 63–5, 103–4, 154, 158–63
  of wage-bill, 58, 63–5, 104, 140, 156 160–3
  of wage-rates, 58, 60–2
Wage control, models of
  differentiated, 108–11
  direct, 51, 59(d), 67, 69–71, 75, 110, 158 182, 188–9, 192
  indirect, 59(d), 67, 69–71, 102–3, 111, 158, 184
  mixed, 59(d), 70–1, 72, 184
  uniform, 108–11
Wage fund, defined, XVIII
Wage growth
  and output of consumer goods, 33–4, 140, 178
  and productivity growth (including statistics), 34–43, 49
  and wage plan targets, 34, 45–8, 181, *see also* Wage
Wage regulation, *see* Wage control
Wagner, U., 212
Ward, B., 99, 212
Weber, H., 218
Wilczek, J., 65, 105, 207, 213
Wiles, P., XVI, 180, 196, 199, 206, 208, 227

Wojciechowska, U., 207, 212–13, 219
Wolowczyk, A., 219
Work norms, 41, 61–2, 104, 186
Wosková, V., 210–11, 217
Wrzaszczyk, T., 220

Year-end rewards, 89–90, 144, 153, 173

Zacharov, V., 216

Zadoia, P., 216–17
Zaleski, E., 197–8
Zamojska, I., 219
Zawislak, A., 220
Zielinski, J., XX, 197, 206, 213, 218–19, 225, 228
Zorza, V., 201
Zsóka, L., 208
Zverev, A., 96, 203, 212